YOU START PHYSICS

**CHRIS JAMES
PHIL BLOOMFIELD**

HODDER AND STOUGHTON
LONDON SYDNEY AUCKLAND TORONTO

About this book...

What is Physics? What do physicists do? How does physics affect your life? You can find out the answers to these questions in this book!

The book is divided into seven chapters. Each one looks at an important part of physics. You can find out about forces, light, waves, magnets and electricity, heat and temperature, atoms and molecules, and the electromagnetic spectrum. Together, these topics make up much of what physics is all about. Every chapter has three easy-to-read sections. The first one has lots of diagrams and interesting information about a topic. The second section contains activities and investigations for you to try. The last section has some questions about the topic. These will refresh your memory about the topic—and get you thinking as well.

After reading this book, you'll find that physics is fascinating—and fun too!

First printed 1986

Copyright © 1986 C. R. James and P. E. Bloomfield

All rights reserved. No part of this publication may be reproduced or transmitted in any form or by any means, electronic or mechanical, including photocopy, recording, or any information storage and retrieval system, without permission in writing from the Copyright Licensing Agency Limited. Further details of such licences (for reprographic reproduction) may be obtained from the Copyright Licensing Agency Limited of 7 Ridgemount Street, London WC1E 7AE.

Typeset in 11/13pt Rockwell by Rowland Phototypesetting Ltd,
Bury St Edmunds, Suffolk
James, C.R.
You start physics.
1. Physics
I. Title II. Bloomfield, P.E.
530 QC23

ISBN 0 340 37157 9
Printed in Hong Kong for
Hodder and Stoughton Ltd, Mill Road, Dunton Green, Sevenoaks, Kent, TN13 2YD, by Colorcraft Ltd

Thanks to...

Photographs: Heather Angel (8 middle, 24, 96 top middle) British Rail (86 bottom left) Clive Barda (44 middle) Bilsom International (48) British Telecom (110 middle, 119 top) E. Bolshaw (22, 26, 57, 61 middle, 81, 85 top, 96 bottom, 108, 116 bottom) Casio (45) Colorsport (6 top right, 9, 38, 83) Corning Ltd (95 top) Dunlop (4 top) Fotocall (123) GEC (58) Griffin and George (5 top, 87, 29 bottom) John Harris (61 top) Inmos (67) Institute of Orthopaedics (111 middle top) JET (103 bottom) Alex von Koettlitz (44 top) Robert McNaught (12 middle and bottom, 86 middle) Media Services (13) A. E. Morgan (25) NASA (6 top middle, 19 bottom middle and bottom right) National Remote Sensing Centre (114 middle right) Philips (110 bottom right, 111 bottom middle, 116 top) Popperfoto (6 bottom right, 8 top, 42 top left) RSPB (80) Royal Greenwich Observatory (11, 12 top, 23 top, 103 middle) Science Photo Library (19 bottom left, 23 bottom, 100, 114 bottom left) Supersport (6 top left and bottom left, 114 top) S. Taylor (46) John Topham (85 middle) UKAEA (101, 117 bottom right) Pat Ward (4 middle) Space Frontiers (18, 19 passim) Cover photographs Phil Holden (top left) Philips (bottom left) Science Photo Library (bottom middle) Supersport (right) N. James (19 top) Artwork: Robert Britton and Eira Reeves

CONTENTS

About this book 2

Forces and movement 4

Light, lenses and mirrors 22

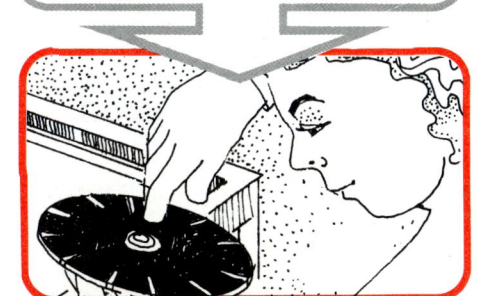

Waves 40

Magnets and electricity 56

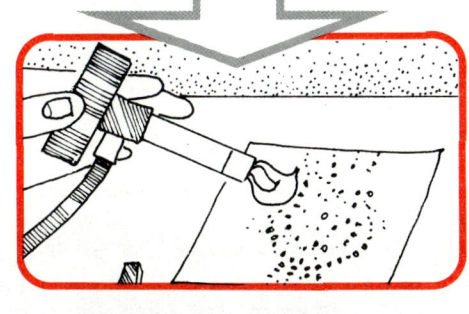

Heat and temperature 78

Atoms and molecules 94

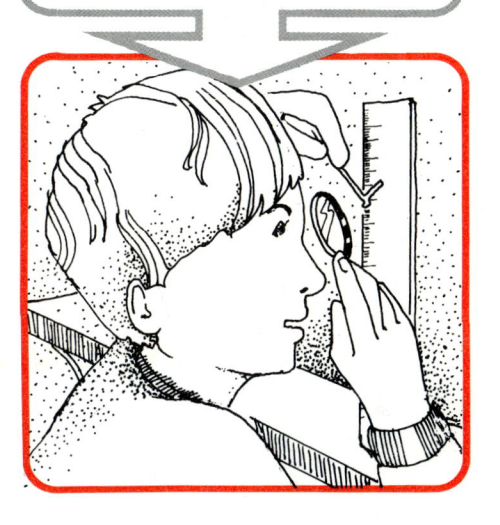

The electro-magnetic spectrum 110

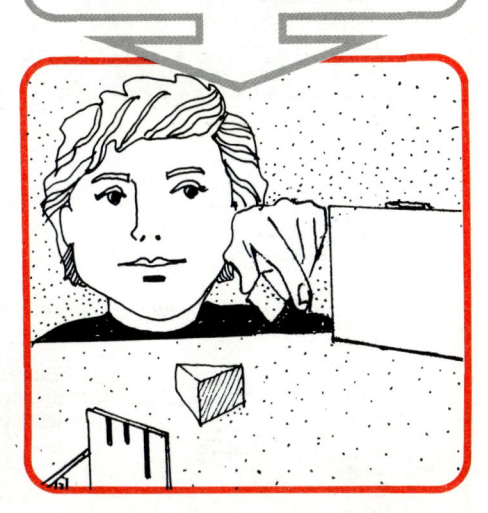

Index 126

PHYSICS TELLS YOU HOW THE UNIVERSE WORKS...

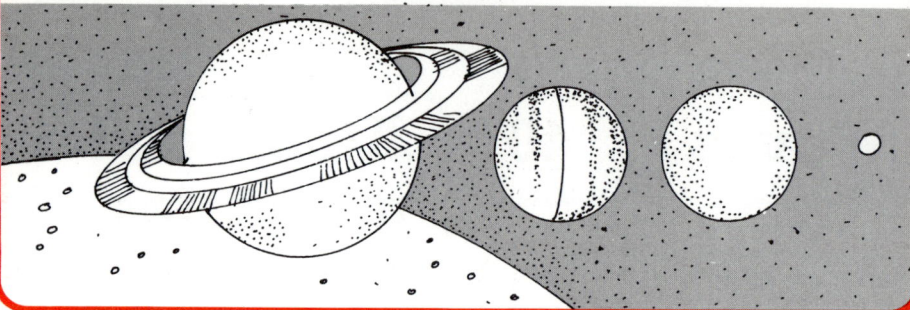

YOU CAN DO PHYSICS!
- Observing
- noting and drawing
- Making measurements
- Predicting—whats going to happen
- Testing—were you right?
- Doing experiments
- Having ideas
- Having theories

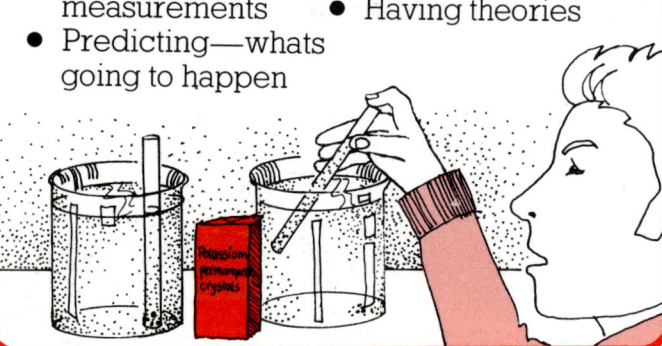

LEARNING PHYSICS WILL HELP YOU UNDERSTAND HOW THINGS WORK
- Computers
- Synthesizers
- Lasers
- Rockets
- Radios
- Televisions
- Magnets

WHAT IS PHYSICS? IT'S ABOUT

WHY DO PHYSICS?

STUDYING PHYSICS WILL TELL YOU ABOUT...

- Electricity
- Heat
- Waves
- Colour
- Movement
- Radioactivity
- Atoms
- Molecules
- Light
- Sound
- Energy

IF YOU LIKE PHYSICS YOU COULD BECOME A...

- Engineer
- Nurse
- Designer
- Architect
- Photographer
- Mechanic
- Astronaut
- Technician
- Electrician

PHYSICS IS ONE WAY YOU COULD LOOK AT THE WORLD...

1 FORCES AND MOVEMENT

KINDS OF FORCES

There are many different kinds of forces. Can you pick out a pull, push, bend, stretch, twist and squeeze in this picture?

Getting moving

It takes a force to make something move. It also takes a force to make something stop.

This golf ball is moving now! This is because of the force from the golf club.

The girl in the picture has changed the direction of the ball. She has given it just enough force to deflect it into the goal. The net will force the ball to stop.

It seems that *if something is standing still it takes a force to get it moving. If it's moving already it takes a force to stop it or change its direction.*

Gravity

You know that when you jump up, you are pulled back down to the Earth. A force pulls you back. It is the force of **gravity**. It is pulling on you now. The pull of gravity comes from the enormous mass of the Earth. You can learn more about gravity on page 16.

Gravity gives you your weight. Your weight is the force that you push with on whatever is supporting you. This could be the ground, a stool, a trampoline, or an aeroplane! What is supporting you as you read this?

If there was no gravity you wouldn't have any weight. An astronaut in deep space, away from all stars would be completely weightless.

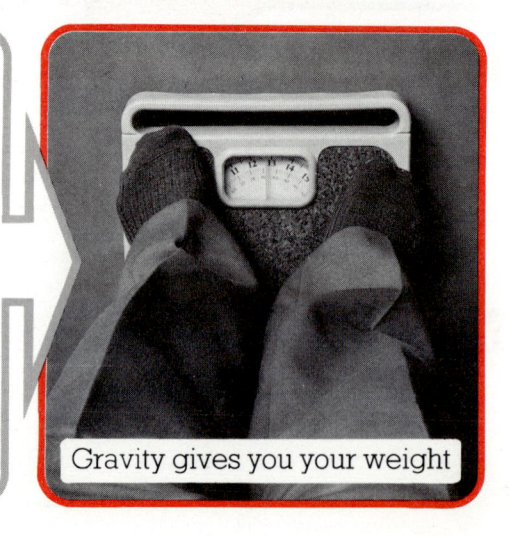

Gravity gives you your weight

Measuring forces

We measure force in **newtons** (N for short). Because weight is a force, we measure weight in newtons, too.

How big is a newton? An average eating apple has a weight of one newton. In other words, the Earth pulls it downwards with a force of one newton.

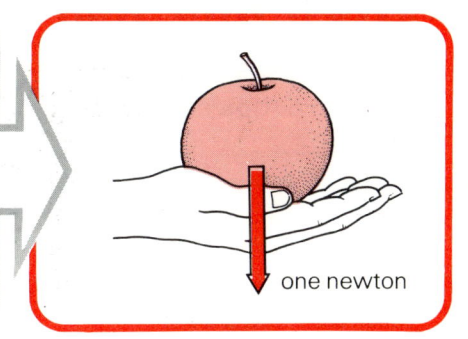

one newton

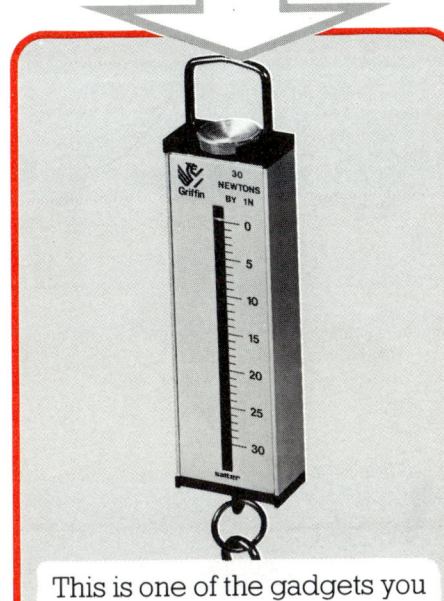

This is one of the gadgets you can use to measure force. It is a **newton-meter**.

The girl is measuring the force needed to pull the door open. It takes about 5 N.

The fisherman is weighing his catch using a newton-meter. 50 N, a beauty!

What is friction?

Friction is a force. It is the force that stops things sliding across each other. Try rubbing your hands together. After a short time they get hot. This is because of the friction between them.

Look at this picture of a science fiction space-ship exploding. After the explosion, the bits of wreckage will go on moving through space at the same speed. This is because there is no air in space to rub against the wreckage as it moves. So, there is no friction to slow it down. The bits of wreckage will only stop or change direction when a force acts on them.

Look at this picture of a girl riding a bike. You might think, if you did not know about how friction slows things down, that one push on the pedal would send this bike to the end of the street. But it doesn't happen. This is because there is friction between the wheels and the ground. This slows the bike down. She has to keep on pedalling to overcome the friction.

5

ACTION AND REACTION

Look at these pictures carefully. You will see that forces come in pairs, an **action** and a **reaction**.

The man fires his gun. The bullet is forced forward (action). He feels a backwards 'kick' from the gun (reaction) against his arm.

The gases and flames shoot downwards out of the rocket (action). The rocket is forced upwards (reaction).

The swimmer pushes the water backwards (action) and she is forced forward (reaction).

Action, reaction and friction

You could not walk along the street without the help of friction! There is friction between the soles of your shoes and the ground. You move forward because you push against the ground (action). The friction is the backwards reaction.

Walking on *frictionless* ground would be impossible. You get an idea of what it might be like when you first go ice-skating. But even here there is some friction between the ice and your skates.

How does the astronaut move about in airless (frictionless) space? He uses small rocket motors!

MOMENTS

What are they?

Try holding this book out at arm's length for a while. You would have to be very strong to keep it out there for more than 10 minutes. But it's easy to hold the book close to your chest. You could probably do *that* for hours!

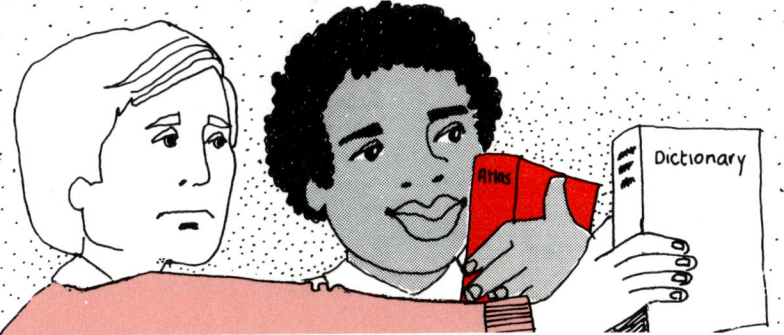

Why should this be? The weight of the book is the same which ever way you hold it. The difference is in the turning effect of the weight. It is much bigger when you hold the book away from you. We call this turning effect the **moment** of the force.

Spanners

You use a spanner to increase the moment of a force. Imagine trying to tighten a nut just using your fingers. You would never get it really tight. You would if you used a spanner, even though you used the same force with your fingers.

Small moment

Larger moment

Very large moment

Levers

You can use a screwdriver to lever the lid off a tin of paint. The moment of the force you use is increased by the length of the screwdriver – the longer the screwdriver, the bigger the moment, the better lever it is. The screwdriver pivots on the edge of the tin. The place where a lever pivots is called the **fulcrum**.

See-saws

A see-saw is an example of moments in action.
The see-saw is balanced. The girl is heavier than the boy but she is sitting nearer the fulcrum. The moment of the boy's weight is exactly equal to the moment of the girl's weight.

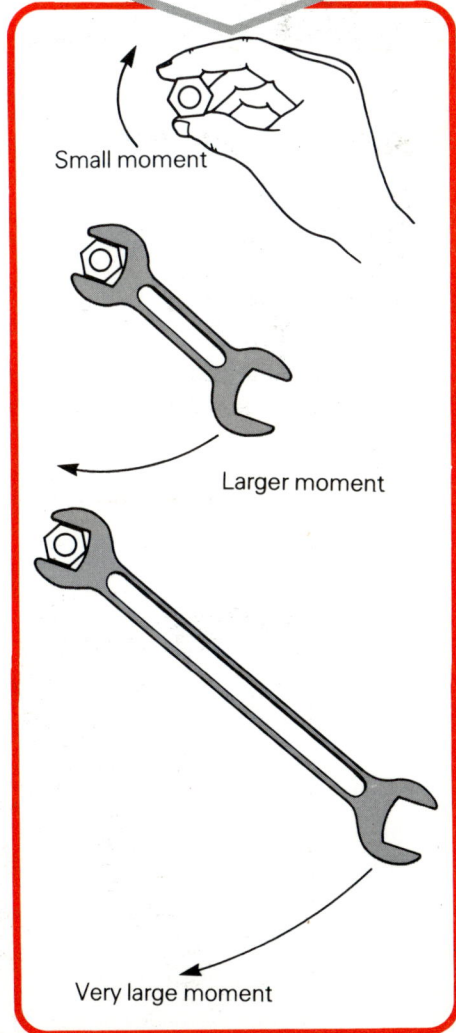

Centre of gravity

Moments are important whenever you want to balance!
Try to balance a metre rule at the 75 cm mark. You can't! The ruler pivots where your finger is. The longer side goes downwards.

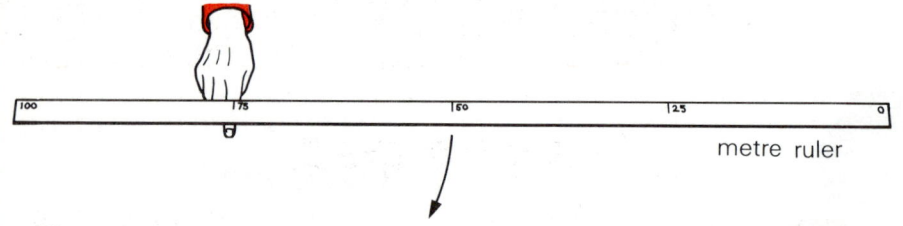

There is only one place where the ruler *will* balance. For a normal metre rule, this will be at the 50 cm mark. This place is called the **centre of gravity** of the ruler. The moments on both sides of it are equal.

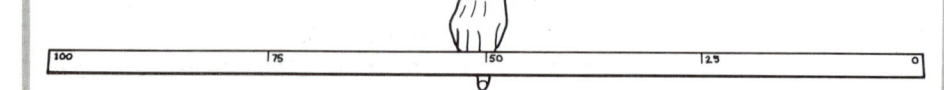

Balance

Look at the picture. The ball's centre of gravity is in the middle. The seal moves so that its nose is exactly underneath the ball's centre of gravity. Only when the seal does this will the ball balance.

Stability

Something's stability is affected by the position of its centre of gravity.
Look at this brick being pushed over.

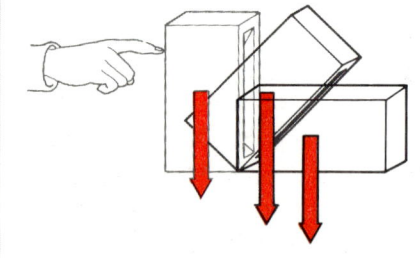

There is an arrow drawn straight down from the centre of gravity. When that line moves outside the base of the brick, the brick topples over.

Look at these two vases of flowers.

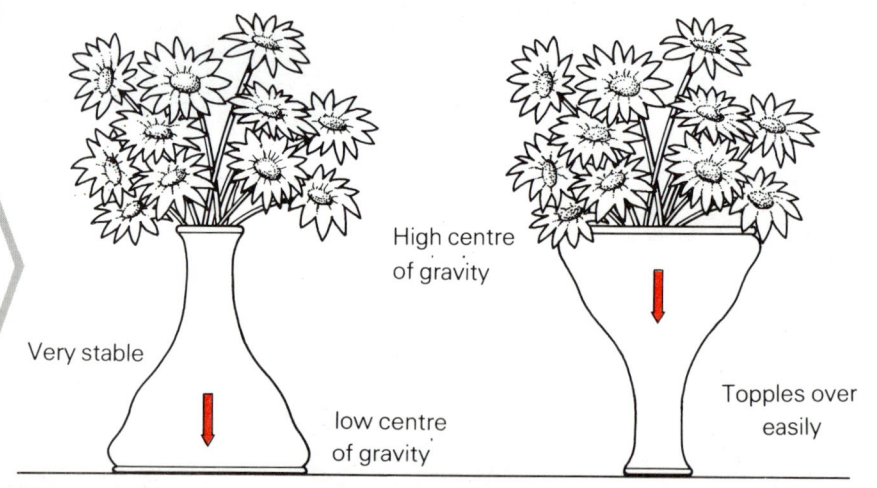

This one has a low centre of gravity and a wide base. It has to be tipped a long way before it topples over.

This vase has a much higher centre of gravity and a narrower base. It is much less stable.

MOVEMENT

This car is moving fast! You can find out just how fast it is going by measuring how far it goes in one second. We call this measurement its **speed**. The speed of the car is the *number of metres it moves in one second*.

Measuring speed

The simplest way of measuring the speed of the car is to measure
1 how far it moves in metres,
2 how long it takes in seconds.
Then divide

$$\frac{\text{distance moved}}{\text{time taken}}$$

This gives you the speed.

Look at this example. A boy cycled to his friend's house. He cycled 400 m and it took him 80 seconds. What was his speed?

$$\text{speed} = \frac{\text{distance moved}}{\text{time taken}}$$

$$= \frac{400 \text{ m}}{80 \text{ s}} = 5 \text{ m/s}$$

Changing speed

To increase the speed of a car the driver presses the accelerator. This increase in speed is called **acceleration**. It's actually the change in speed during one second. Imagine that the car increased speed from 5 m/s to 8 m/s in a second. Its acceleration would be 3 m/s per second. This is usually written as 3 m/s^2 (say 3 metres per second *squared*). You can try an interesting way of measuring acceleration in the Activity on page 17.

Speed or velocity

Physicists often use the word **velocity** instead of speed. These words mean almost the same thing but velocity includes direction.

So you might say 'the car's speed is 40 m/s.' or 'the car's velocity is 40 m/s *towards the finishing line.*'

Speed of falling objects

When an object falls to the ground it accelerates. This is because it is attracted by gravity. The acceleration is called the **acceleration due to gravity**. (You can measure this in the Activity on page 16). Is the acceleration due to gravity the same for everything? Yes! – but it's hard to believe it. Look at this experiment.

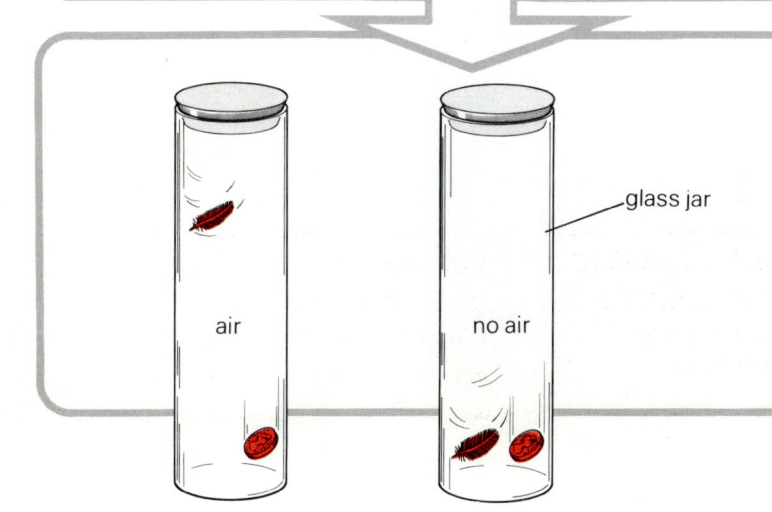

If you drop a feather and a 10p piece at the same time, they won't fall together. This is because air resistance slows the feather more than the coin. You could see what happens without air by watching objects fall in a vacuum. This time you'd see that they do fall with the same acceleration. There isn't any air resistance on the Moon. One of the astronauts who went to the moon tried dropping a hammer and a feather at the same time. They landed together!

ASTRONOMY

You and the Universe

Because of gravity, everything in the Universe is attracted to everything else. Even the person sitting next to you is attracting you – gravitationally speaking of course! But you cannot feel anything because his or her mass is so small. The size of the attraction depends on the masses of the things doing the attracting and how far apart they are. Things with more mass attract more. The attraction is less when they are further apart. The Earth has a large mass. It attracts you a lot, but you attract the Earth too.

Gravity keeps the Moon close to the Earth. It also holds the Earth and the other planets near the Sun.

The Earth and the Sun

The Earth moves round the Sun. Once round takes one year. So, we've all been round the Sun once since this time last year! This journey round the Sun is called the Earth's orbit.

The Earth spins about an imaginary line which runs from the North Pole to the South Pole. One complete spin around takes a day. So, we've all spun round once since this time yesterday!

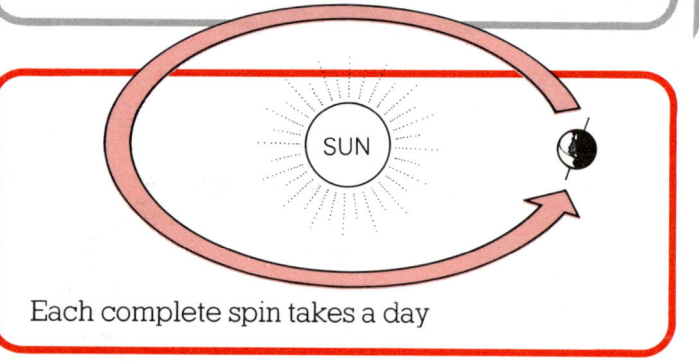

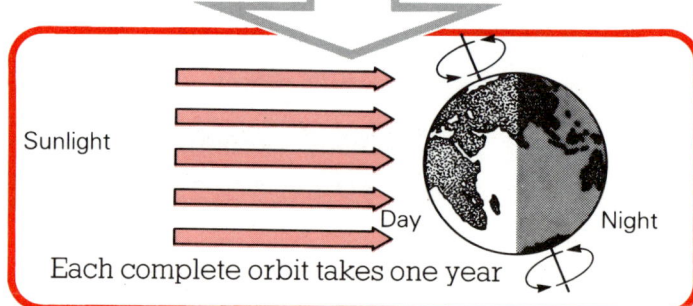

Each complete spin takes a day

Each complete orbit takes one year

The planets

There are nine planets orbitting the Sun. Together they make up the **Solar System**. The Earth is the third closest planet to the Sun.

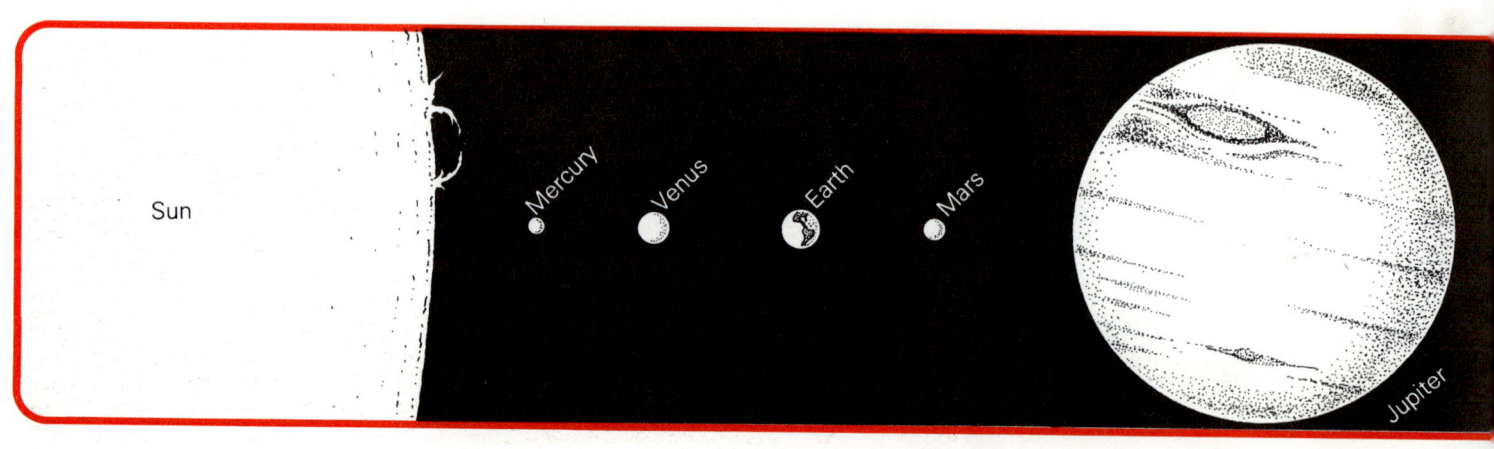

Sun   Mercury   Venus   Earth   Mars   Jupiter

10

The Moon

Several planets have moons orbitting around them. Mars has two, Uranus has five and Jupiter has 14.

The Earth only has one. We call ours *the* Moon. It orbits the Earth once every 28 days. The Moon also spins round once every 28 days. Because of this the same side faces us all the time.

The surface of the Moon is very different from the Earth. There is no water, no atmosphere and no life. Unlike the Earth, the Moon hasn't changed for millions of years. When you look at the Moon you can see darker areas called **seas**. These seas aren't wet, they contain the solid lava from volcanoes. There are also many craters. These are places where lumps of rock have crashed onto the surface.

The Moon's mass is about 1/80 of the Earth's mass. This picture gives you an idea of the Moon's size. The pull of gravity between the Earth and Moon is strong enough to move the water in the seas on the Earth's surface. This causes the tide to go in and out.

How many times could you fit the UK onto the moon's surface?

The Moon never looks the same!

The Moon doesn't give off any light of its own. You can only see the parts of it that reflect sunlight to the Earth. The positions of the Sun, the Moon and the Earth are changing all the time. So, the amount of the Moon you can see changes as well. Because of this, the Moon's shape seems to change.

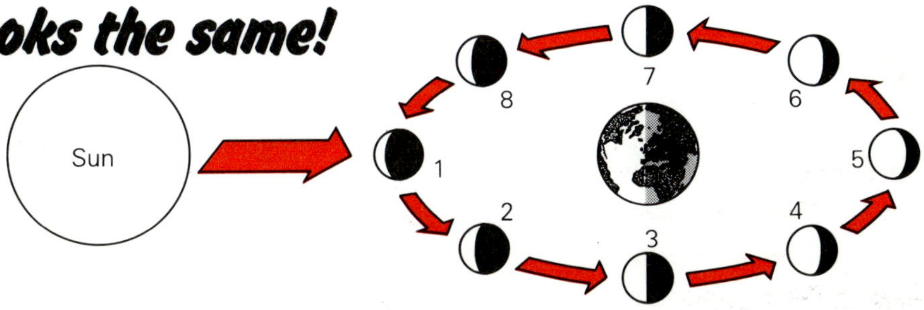

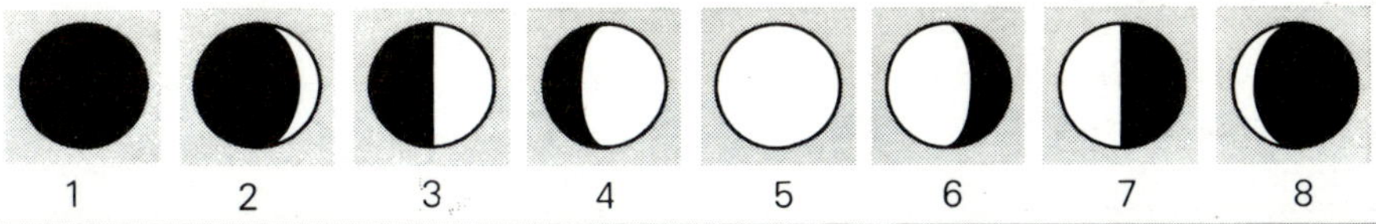

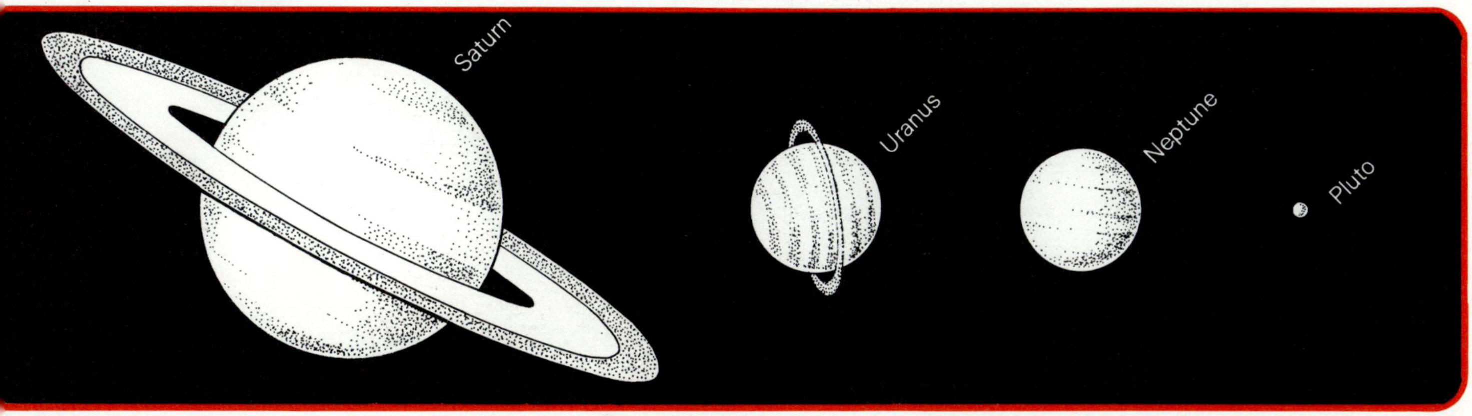

Comets

Comets are very small objects which orbit the Sun. Look at this picture of the famous Halley's comet. It clearly shows the head and the tail. Astronomers think that comets are made from dust mixed up with frozen gas. They are rather like 'dirty snowballs'. As the comet gets near the Sun, the frozen gases in the head of the comet warm up. They evaporate to become the tail.

Halley's comet

Meteors

Have you ever seen a 'shooting star'? It was probably a meteor. **Meteors** are tiny bits of dust entering the Earth's atmosphere. A meteor burns up very brightly. The heat comes from the friction between the meteor and the air.

Meteorites are larger lumps of rock or iron. They actually reach the surface of the Earth without burning-up completely. Fortunately, they are very rare!

This crater in Arizona, USA, was made by a meteorite

The Milky Way

The Sun is part of a collection of 1 000 000 million other stars called the **Galaxy**. *All* the stars you can see are in our galaxy. The main part of the galaxy shows up as a band of stars across the middle of the sky. This is called the Milky Way. You can see it on a very dark clear night. Try looking at it through binoculars. You will see that the milkiness comes from millions of faint stars.

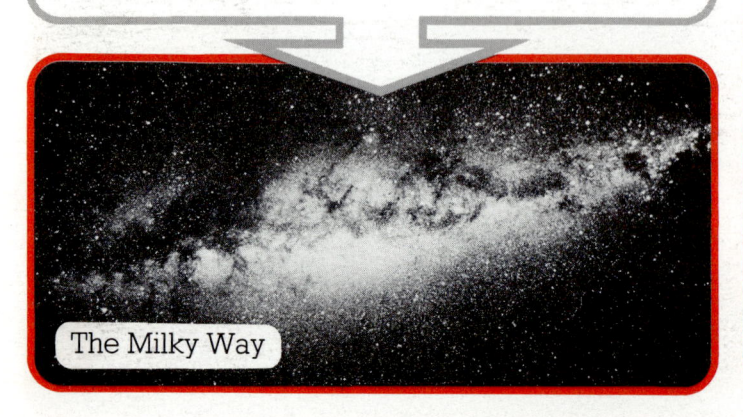

The Milky Way

you are here

Our galaxy seen from the top

you are here

Our galaxy seen from the side

The stars

The distances between stars are enormous. To go to the nearest star in a jumbo jet (if it could fly in space!) would take about 4 million years! If you could to visit the furthest star in the same way you could take up to 400 000 million years. It must be a long way!

Many of the nearest stars are known to us by their positions. These hardly change from year to year. The patterns the stars make are known as **constellations**. One of the better known ones is called the Plough.

The two stars on the right side of the Plough point to the Pole Star. This is sometimes called the North Star. It is directly above the North Pole of the Earth. If you can find this star at night you will know which direction North is.

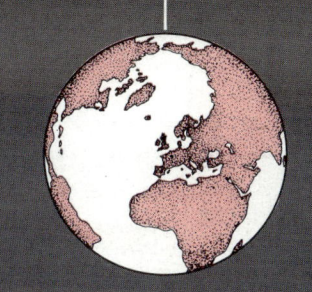

Galaxies

Our galaxy isn't the only one. The Universe is mostly empty space, but within that space are many other galaxies. This picture shows another galaxy in our 'local' group.

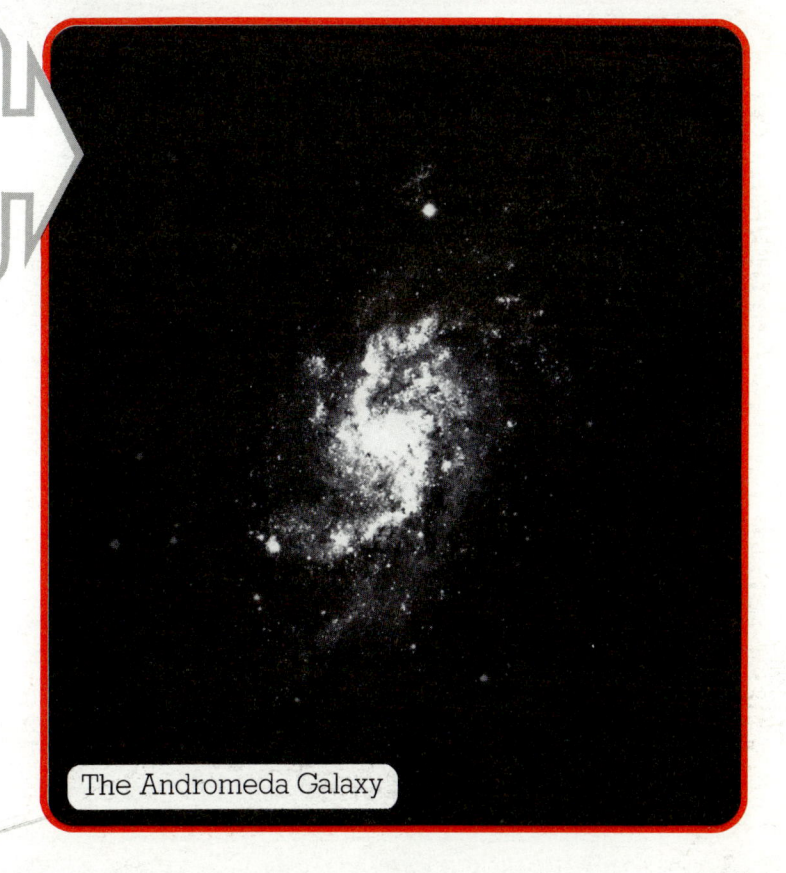

The Andromeda Galaxy

MEASURING SPEED WITH TICKER TAPE

You can use ticker tape and a ticker timer to measure speed.

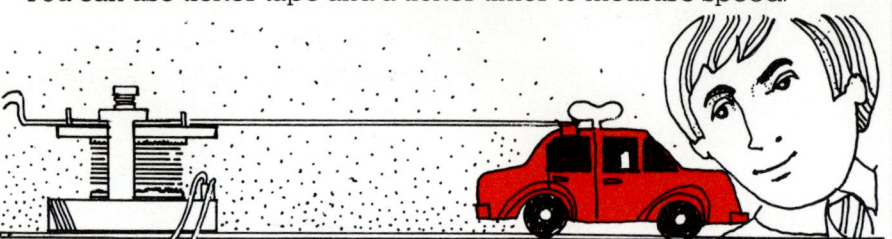

You will need
- toy car
- scissors
- sticky tape
- ticker timer
- carbon paper disc
- 1 m ticker tape
- 12 V a.c. power supply

The ticker timer makes 50 dots a second on the tape. When the car moves, it pulls the tape through the ticker timer. This makes a recording of the movement. You can use this recording to find out the car's speed.

1 Wind the car up and let it run across the bench. Estimate its speed in cm/s.

2 Arrange the equipment to look like the picture at the top of the page. Connect the ticker timer to the power supply. Make sure the tape won't get caught when the car pulls it through the timer.

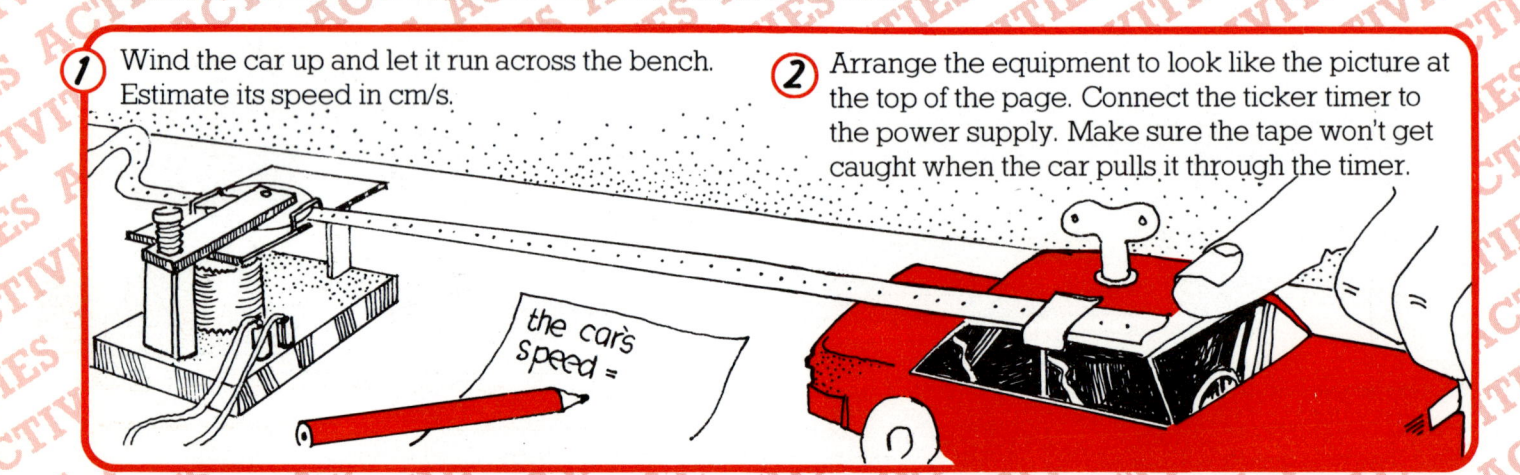

3 Wind up the car. Turn on the power supply. Let the car go. Stop the car when it's pulled the tape through. Turn off the power supply. Take the tape off the car.

4 Look at the tape. Remember:
(a) 10 spaces are worth $\frac{1}{5}$ second.
(b) in $\frac{1}{5}$ s the car went 10 spaces.

(c) the speed (in cm/s) = $\dfrac{\text{how far the car went}}{\text{the time it took}}$

So the car's speed = $\dfrac{\text{length of 10 spaces in cm}}{\frac{1}{5} \text{s}}$

Work out the speed again using two other sections of tape.

Answer these questions

1. What were your three measurements of the car's speed?
2. What was the average of the three speed measurements you made?
3. Compare your estimate of the car's speed with the average speed.
4. How many spaces on the tape make up $\frac{1}{10}$ second?

MEASURING SPEED CHANGES

In this Activity you are going to find out how the speed of a trolley changes as it rolls down a slope. You are going to use the ticker tape method. (Look at page 14.)

You will need
- trolley • scissors
- ticker timer • 2 clamps
- carbon paper disc
- 2 m ticker tape
- runway • 2 stands
- 12 V a.c. power supply

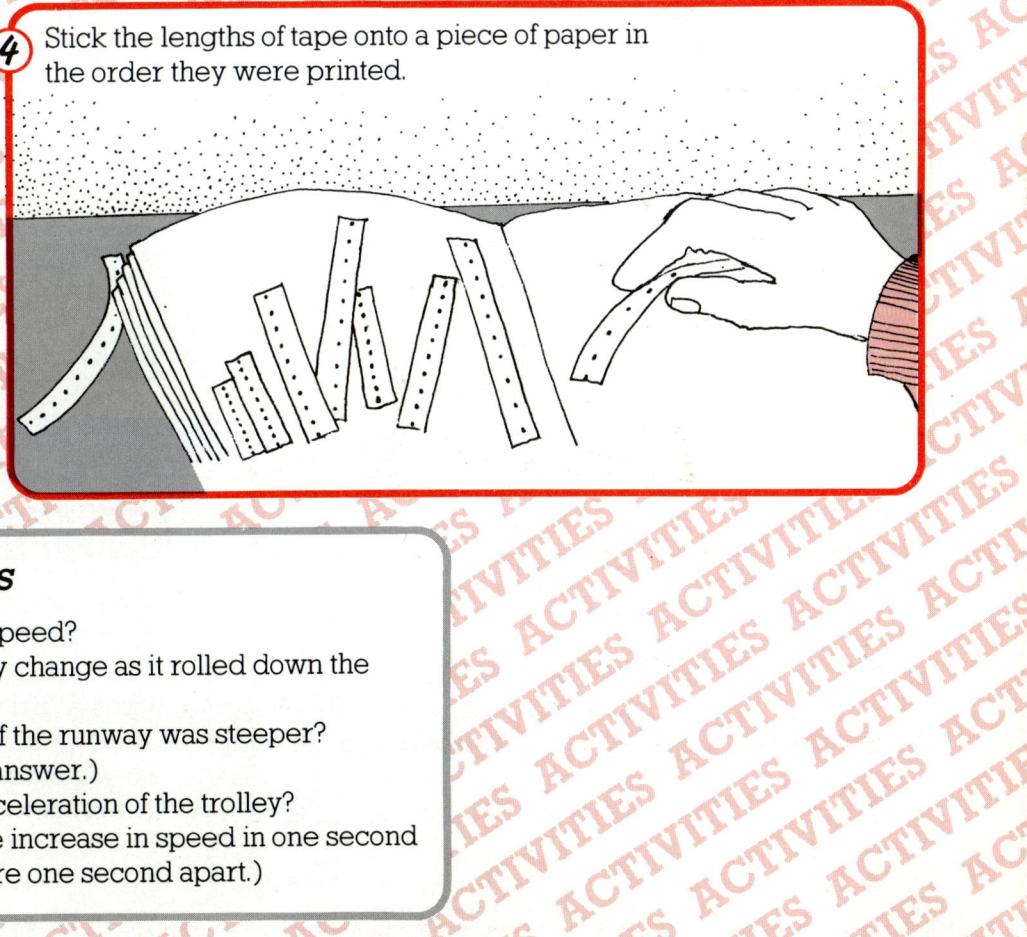

1 Fasten one end of the tape to the trolley. Feed the other end through the timer. Make a slope with the runway and the clamp stands or some books.

2 Switch on the ticker timer. Let the trolley roll down the slope. Make sure the tape doesn't get caught as it's pulled through.

3 Cut the tape into ⅕ second lengths. (That's 10 spaces.)

4 Stick the lengths of tape onto a piece of paper in the order they were printed.

Answer these questions

1. What was the trolley's highest speed?
2. How did the speed of the trolley change as it rolled down the slope?
3. How would the results change if the runway was steeper? (You could try it to check your answer.)
4. How could you work out the acceleration of the trolley? (Remember: acceleration is the increase in speed in one second *and* the first and sixth lengths are one second apart.)

WHAT IS THE ACCELERATION DUE TO GRAVITY?

In this Activity you are going to use the ticker tape method to measure the acceleration due to gravity. You are going to make a ticker tape recording of a mass as it falls.

You will need
- ticker timer
- carbon paper disc
- 12 V a.c. power supply
- 4 m ticker tape
- a 100 g mass

1 Feed the tape through the timer. Stick the end on to the mass.

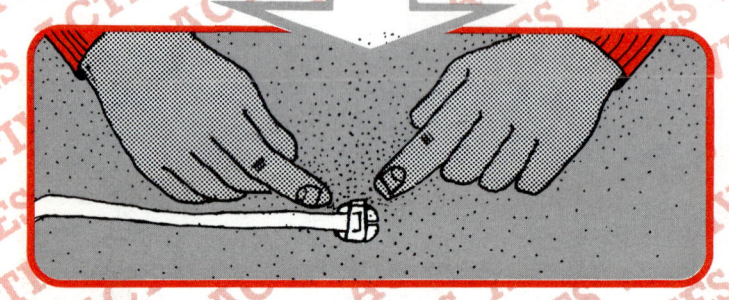

2 Switch on the ticker timer. Drop the mass. Switch off the ticker timer.

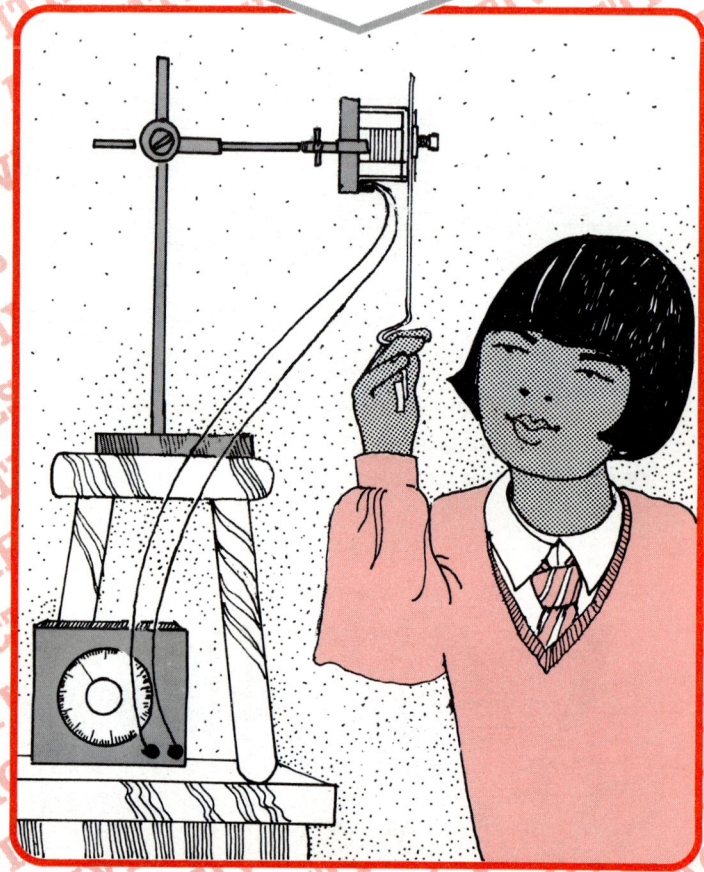

3 Cut the tape up into 1/5 s pieces. Look at two neighbouring strips from the middle of the tape. Work out the speed for each of these strips.

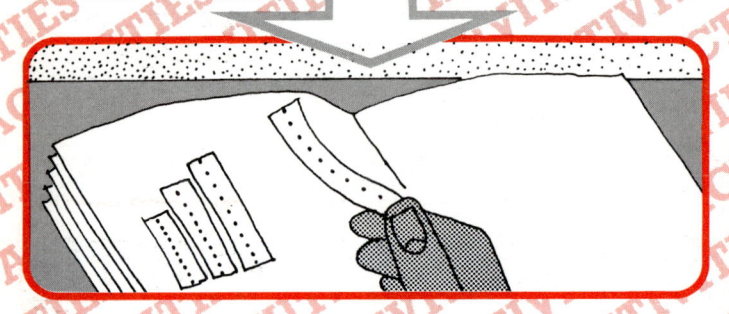

4 From these two strips you can see how the speed has changed in 1/5 s. (1/5 s is the time between the two strips.) So, you can work out the acceleration:

$$\text{Acceleration} = \frac{\text{change in speed}}{\text{time of change}}$$

$$= \frac{\left(\begin{array}{c}\text{speed of strip 1}\\ \text{in cm/s}\end{array}\right) - \left(\begin{array}{c}\text{speed of strip 2}\\ \text{in cm/s}\end{array}\right)}{1/5 \text{ second}}$$

$$= \text{-----} \text{ cm/s}^2$$

Answer these questions

1. What was your value for the acceleration due to gravity?
2. Collect the results of other people in your class. What was the average result?
3. Would you have got a different result if you had used a bigger mass? Find out by repeating the experiment with two masses taped together.
4. How could you improve the method used in this Activity?

FINDING THE CENTRE OF GRAVITY OF A CARDBOARD ISLAND

When something hangs freely, its centre of gravity must be directly under the point where it is hanging. You can use this fact to find the centre of gravity of a cutout shape of an island.

You will need

- cardboard
- scissors
- a 100 g mass
- cork
- map of an island
- pin
- tracing paper
- thread
- clamp and stand

1 Trace the map of the island onto cardboard. Cut out the shape.

2 Fix the cork in the clamp and stand. Stick the map onto the cork with the pin. Make sure the map swings freely.

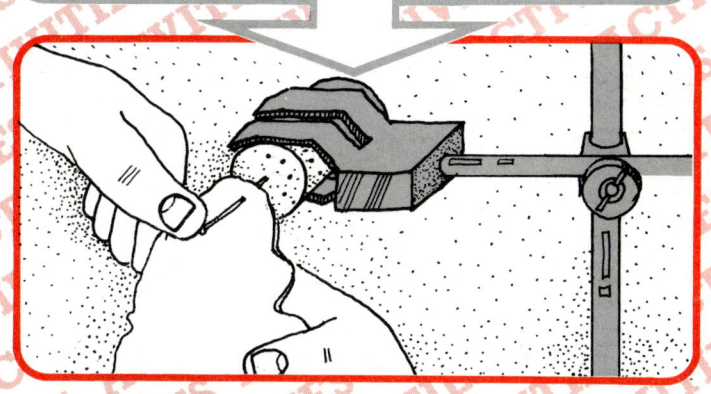

3 Tie the thread onto the mass. Tie the other end to the pin.

4 Mark a line on the card to show where the thread is hanging. Pin the shape to the cork in two more places. Mark the line of the thread each time. The centre of gravity is at the meeting place of the three lines.

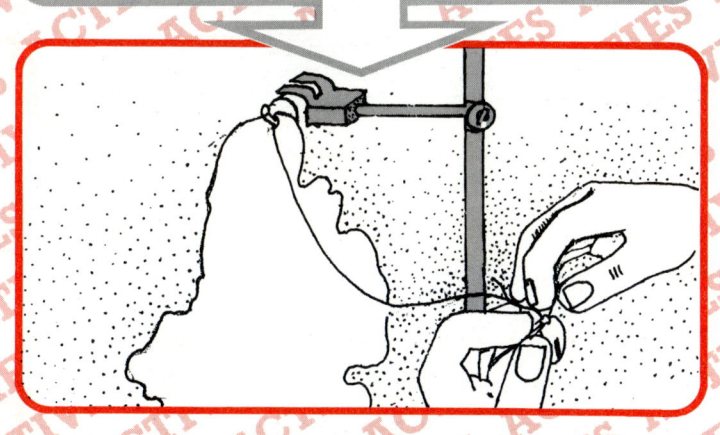

Answer these questions

1. Does your shape balance on a pin at the centre of gravity?
2. Would drawing two lines have told you where the centre of gravity was? If so, what was the reason for drawing three lines?
3. Whenever any shape hangs freely, where must its centre of gravity be?

STARTING ASTRONOMY

This Activity will show you how to pick out some of the more interesting sights in the night sky. You can see most of them with the naked eye. But if you can borrow a pair of binoculars, so much the better.

The Moon

The Moon always shows us this face.

Make a list of the features on the photograph. When you look at the Moon, tick them off as you find them. The seas and rays are most clear when the Moon is full. It is better to look for the craters at other times. Draw what you see.

Look at the Moon on one evening. Look at it again at the same time on the following evening. How does it change? Can you explain why?

Features labelled: Sea of Showers, Aristotle, Copernicus, Sea of Tranquility, Kepler, Sea of Humours, Theophilus, Tycho, Ptolemy

Stars

Stars make patterns called **constellations**. You can see them on any clear night. Pick a night when there is no Moon, if possible. Try to stay out for a while so that your eyes get used to the dark. Make notes about what you saw, when you go back indoors. Here are some constellations to look out for.

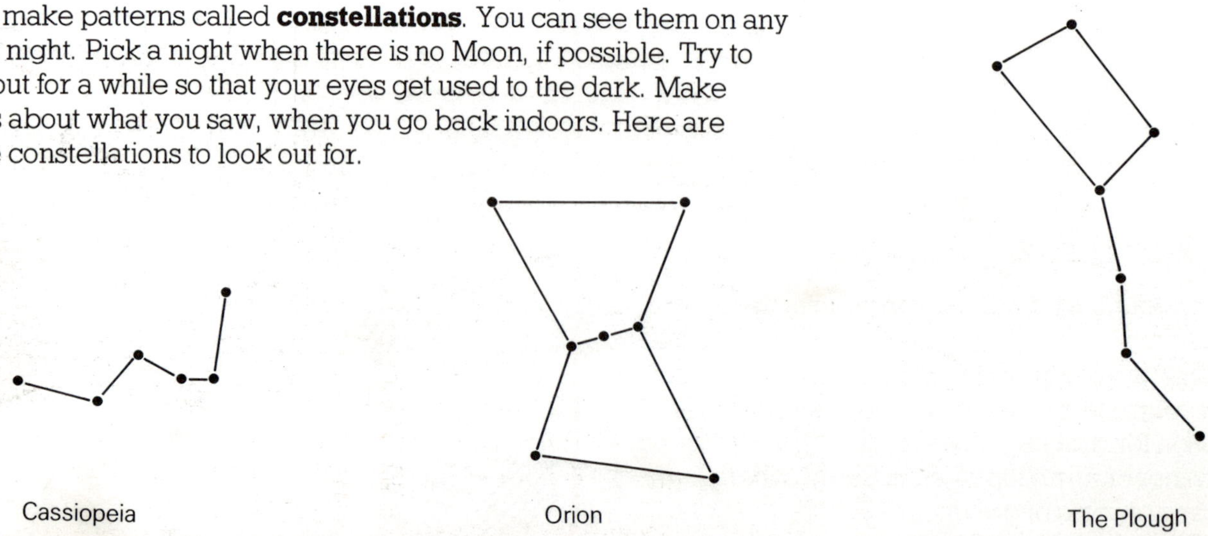

Cassiopeia   Orion   The Plough

Shooting stars

Meteors are sometimes called shooting stars.
You can see meteors on any dark night. But there are certain days in the year when you might see up to 40 meteors per hour! These **meteor showers** are named after the constellations that the meteors *appear* to come from. For example the Orionids seem to come from Orion. Look for the showers for several nights before and after these dates. Make a note of what you see. Count how many meteors you see and try to pin point the place in the sky that they seem to come from.

A meteor

METEOR SHOWER	WHEN YOU CAN SEE IT
PERSEIDS	AUGUST 11
ORIONIDS	OCTOBER 20
TAURIDS	NOVEMBER 7
GEMINIDS	DECEMBER 13
URSIDS	DECEMBER 22
QUADRANTIDS	JANUARY 3
LYRIDS	APRIL 21

Planets

It is impossible in this book to tell you *exactly* where to look to find a planet. This is because they move. In fact, the word 'planet' means wandering star, although planets aren't, in fact, stars! These suggestions will help you find the planets.

Look in the clear night along the *path* that the Sun would take during the day at that time of year. If you are not sure of the Sun's path, check it during the day before.
Look for bright objects that don't 'twinkle' the way stars do. Check from night to night to see if the planet's position changes against the fixed background of stars.
Try to look at the planet through binoculars. Stars always look like tiny specks of light. Planets are larger and may be disc shaped. Look out for star and planet maps published in some newspapers on the first day of the month.

Venus, Mars, Jupiter and Saturn are the planets you can see most easily. Look at them on several nights. Describe what they look like. Note how their positions change in the sky.

Venus is often the brightest object in the sky – apart from the Sun and Moon of course!

Mars is an orange-red colour. Through binoculars, it looks like a disc. You might spot the ice caps at its poles.

Jupiter is the next brightest after Venus. You can see some of its moons if you use binoculars.

Saturn is the ringed planet. You need fairly good binoculars to see the rings.

19

EXERCISES

1. A pull is a force, and so is a push. How many more kinds of forces can you think of?

2. How do we measure forces? Explain how you could measure the force needed to:
   (a) stretch an elastic band,
   (b) break a piece of plasticine,
   (c) lift an apple.

3. What is friction? Explain why a spacecraft glows red-hot when it re-enters the Earth's atmosphere.

4. Imagine that you were in the middle of a very large totally frictionless ice-rink. You can't walk, or skate, because there's no friction. How would you get moving?

5. A man tried to get out of a boat that he had just rowed to the side of a lake. He stepped towards the bank, but the boat moved backwards and he fell in the water. What had he forgotten about action and reaction?

6. A helicopter, taking off, is forced upwards. What is the action and what is the reaction?

7. *True or false*? If false, explain why!
   (a) If you balance a ruler on your finger, then the ruler's centre of gravity is just above your finger.
   (b) Fortunately, meteorites always land in the sea.
   (c) There is no friction acting on a skydiver as she descends to Earth.
   (d) The velocity of a car is the distance it moves in a second.
   (e) A ball thrown upwards accelerates but a ball dropped downwards doesn't.

8. Match up the separate halves of these sentences.

A longer spanner means that	make it stable.
Feathers fall more slowly than coins	has a fulcrum in the middle.
A pair of scissors	you can use a bigger moment.
To calculate the speed of a car you need to	because of air resistance.
A bus has a low centre of gravity to	measure the distance it moved and divide it by the time it took.

9 Fill in the missing words:

To measure ...., you take two measurements of ...., take the smaller one away from the bigger one and divide that answer by the .... interval between them.
Choose from: time, acceleration and speed.

10 Copy out and complete this list.

We measure force in newtons
        speed in ....
        acceleration in ....
        distance in ....
        time in ....
        friction in ....
        acceleration due to gravity in ....

11 A girl jogged around a 1500 m circuit. It took her 8 minutes and 20 seconds to complete the distance. What was her speed?

12 Explain the difference between a planet and a moon.

13 What is the nearest star to Earth?

14 Find a book on astronomy in your library. Find out the names of some of the craters, seas and mountains that you can see on the Moon. Over 600 places on the Moon have got names.

15 Voyager I was a spacecraft sent to fly past Jupiter and Saturn. It has now left the Solar System, and is travelling through the galaxy. This plaque is fixed to the side of the spacecraft. It's meant to tell whoever finds it about life on Earth.
   Imagine that you live on a distant planet and have just found the plaque. Try to explain what it means. Remember, you know nothing at all about Earth.

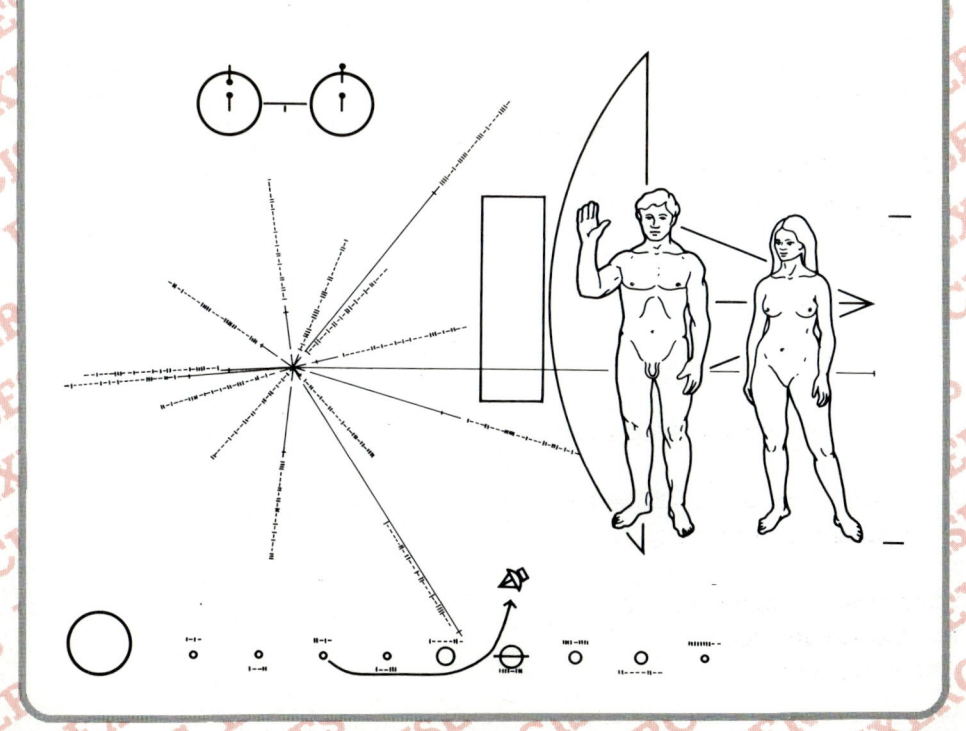

2 LIGHT LENSES AND MIRRORS

WHAT IS LIGHT?

In this chapter you are going to find out some interesting things about light.

WHAT HAPPENS WHEN LIGHT PASSES THROUGH A GLASS OF WATER?

HOW DO MIRRORS REFLECT LIGHT?

WHY DOES THIS WRITING LOOK SO BIG?

HOW DOES LIGHT TRAVEL THROUGH THE AIR?

Shadows

If you shine a torch into the sky on a dark night you will see the clear beam of light. It has straight edges. This tells you that light moves in straight lines. These are called **light rays**. You can easily block the path of a light ray. This makes a shadow behind the block.

The sunlight can't reach the ground where a shadow is. Why not?

Shadow pictures

Have you ever tried making shadow pictures on the wall? The shadow is sharp and black if the lamp is small. If the lamp is large though, the shadow is grey with blurred edges. This is how this happens.

The black lines with arrows on are two of the rays of light that form the edge of the shadow. When the light comes through a small hole, the light rays can't reach any part of the shadow. In the picture only two rays are drawn, to make the diagram clear. Millions more rays could be drawn, but the result would be the same.

When the box is removed from the lamp, the shadow becomes grey and blurred. No light from the bulb reaches the dark part in the middle of the shadow. Some light from at least one part of the bulb reaches the grey part. For example, light from the place marked B on the bulb cannot reach the fly. The light from A can, though!

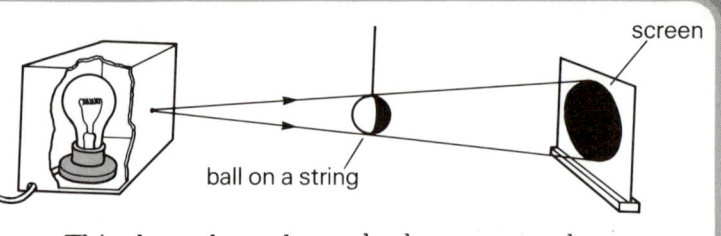

This shows how sharp shadows are made

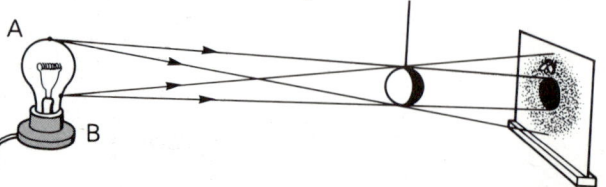

This shows how blurred shadows are made

22

Eclipses

An **eclipse** of the Sun is a shadow on a grand scale. It happens when the Moon comes in between the Earth and the Sun. This blocks off the light coming to us.

How the Sun appears when this happens depends on where you are when the shadow moves across the Earth.

If you are in the dark centre part of the shadow, you will see a total eclipse of the Sun like this photograph.

A total eclipse

The Moon has completely covered the Sun. All you can see is the **corona** around the Sun. The corona is made of hot gas. It is not visible at other times.

It is dangerous to look directly at the sun

If you are in the grey fuzzy part of the shadow then there will be a partial eclipse. It will be something like this photograph.

The next total eclipse of the Sun in Britain is on August 11th 1999.

A partial eclipse

23

REFLECTION

Things like the Sun and a light bulb give out light of their own. Most other things, such as this book, the Moon and you, do not give out light. You only see this book because of the light that comes from the Sun or a lamp. The light **reflects** into your eyes from the surface of the book.

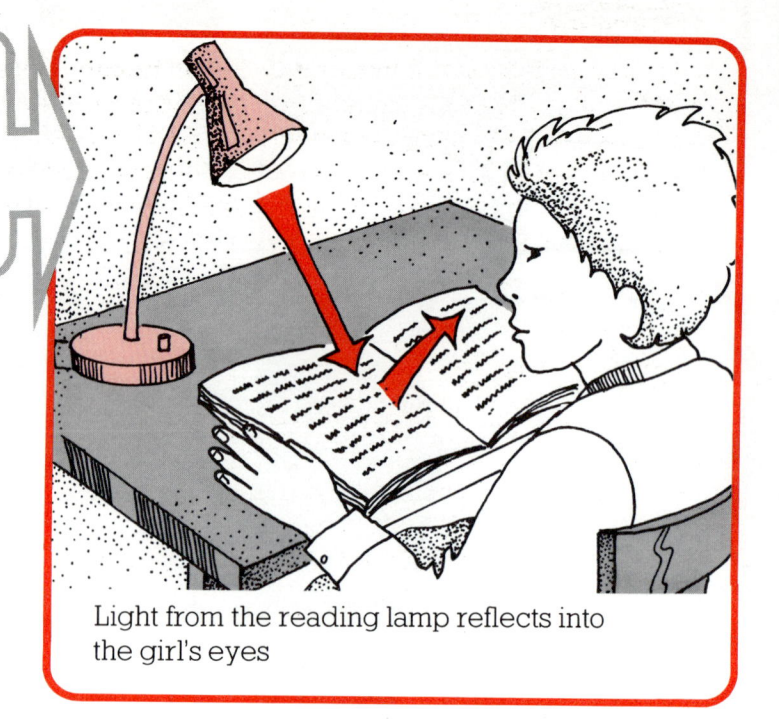

Light from the reading lamp reflects into the girl's eyes

Reflecting light

This picture shows light rays reflecting from a mirror. The **normal line** is a line at right angles to the mirror. The incoming ray and the reflected ray make the same angle with the normal line. This is true for reflection from any surface. (You can find out more about this in the Activity on page 33.)

When light hits a rough surface it is reflected in all directions. Even so, the angles made by each incoming and reflected ray are matched.

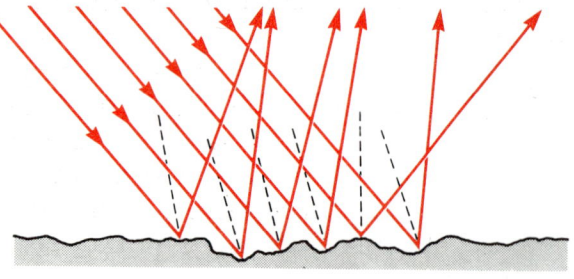

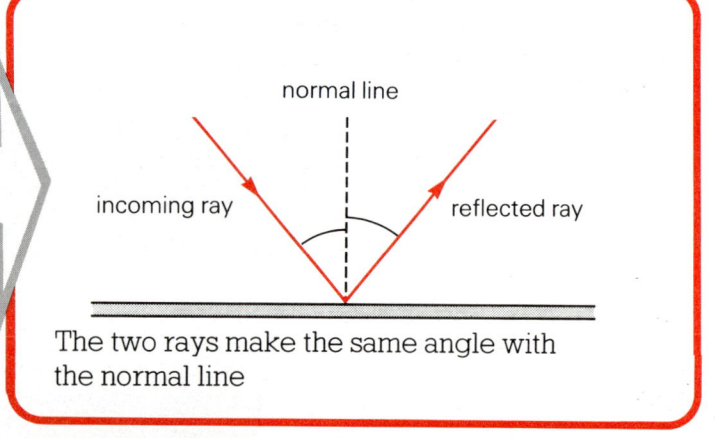

The two rays make the same angle with the normal line

Reflections from a smooth surfaces are regular. You can see a clear reflection when the surface is smooth.

This smooth water gives a good reflection

Images

The clear reflection in a mirror is called an **image**. You and your mirror image aren't the same. If you wink your right eye, your image winks its left eye! Your image is turned around sideways. We say that the image is **laterally inverted**. This ambulance has a strange word on the front. Drivers know what's following them when they read this word in their driving mirrors!

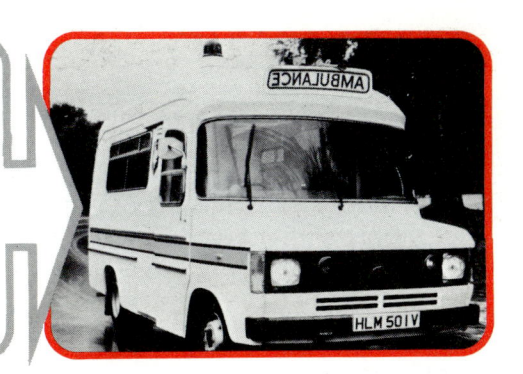

Periscopes

The girl in the picture can see over the heads of the people in front. She is using a simple periscope. It has two mirrors. Light reflects into her eyes from each mirror in turn. The image she sees is the right way round. This is because *both* mirrors laterally invert the image.

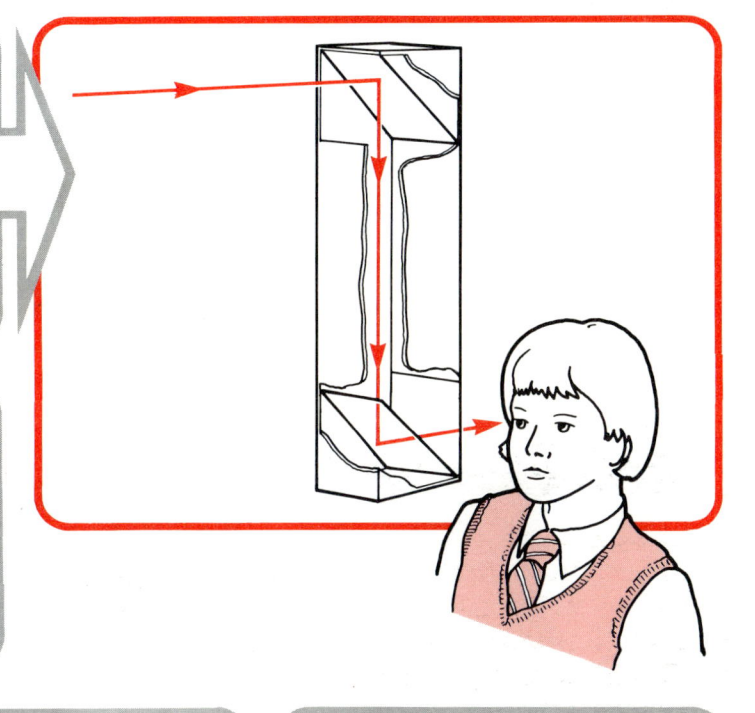

Curved mirrors

Mirrors that curve outwards are called **convex mirrors**. The *back* of a spoon is a good example. Mirrors that curve inwards are called **concave mirrors**. The *front* of a spoon acts as a concave mirror. Curved mirrors can be very useful.

Convex mirrors

Convex mirrors are often used as driving mirrors. They give a better view. You can see more in a convex mirror than you can in a flat mirror of the same size. Convex mirrors are used in other places for the same reason. Have you noticed them in shops and half way up the stairs in double-decker buses?

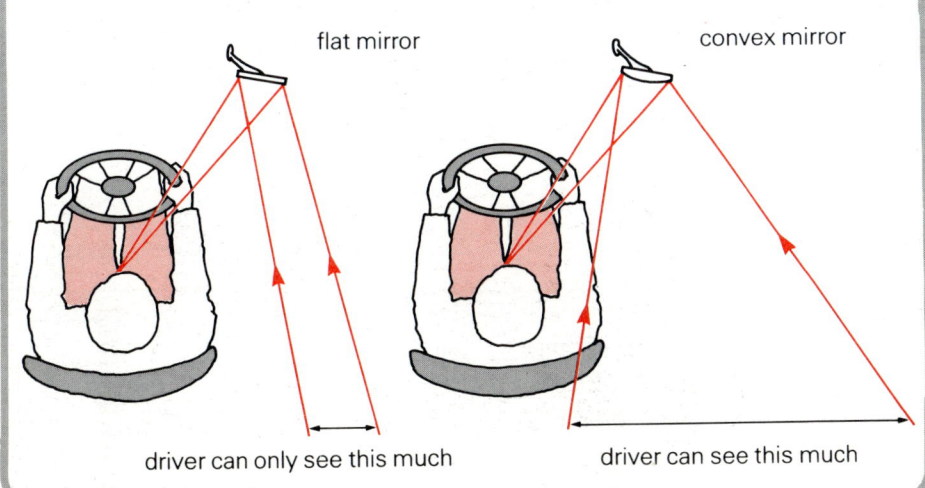

flat mirror   convex mirror

driver can only see this much   driver can see this much

Concave mirrors

If you hold a concave mirror close to your face you will see an image much larger than life. This can be quite useful. Dentists use them to examine your teeth. These mirrors are sometimes used for shaving mirrors, or for putting on make-up more accurately!

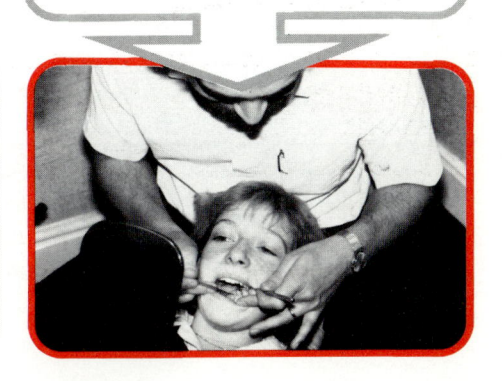

25

REFRACTION

What is it?

If you put a straight stick half in water it looks as if it is bent at the water's surface. It isn't of course! The light from the part of the stick under the water changes direction as it crosses the water's surface. The bending of light in this way is called **refraction**.

Refraction makes the stick look bent

How does it happen?

Refraction happens with glass as well as water. The ray of light in this drawing changes direction twice as it goes through the glass block. What happens is this. When light goes from something thin like air, to something thick like water, it slows down. If the ray hits the glass block at an angle it will change direction. As the light leaves the glass block the reverse happens. It speeds up. This moves the ray back to its original direction.

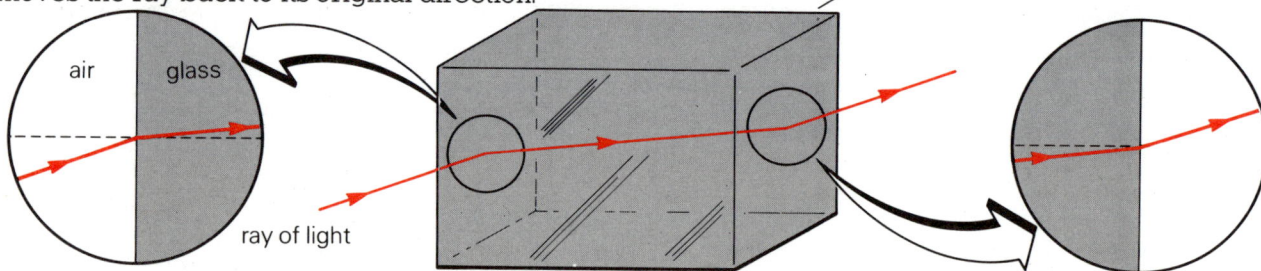

On the way into the glass the ray bends towards the normal line.

On the way out of the block the ray bends away from the normal line.

Mirages

If you look along a road on a hot day you sometimes see pools of water. They aren't there really. It just *seems* so. This is called a **mirage**. It is caused by refraction.

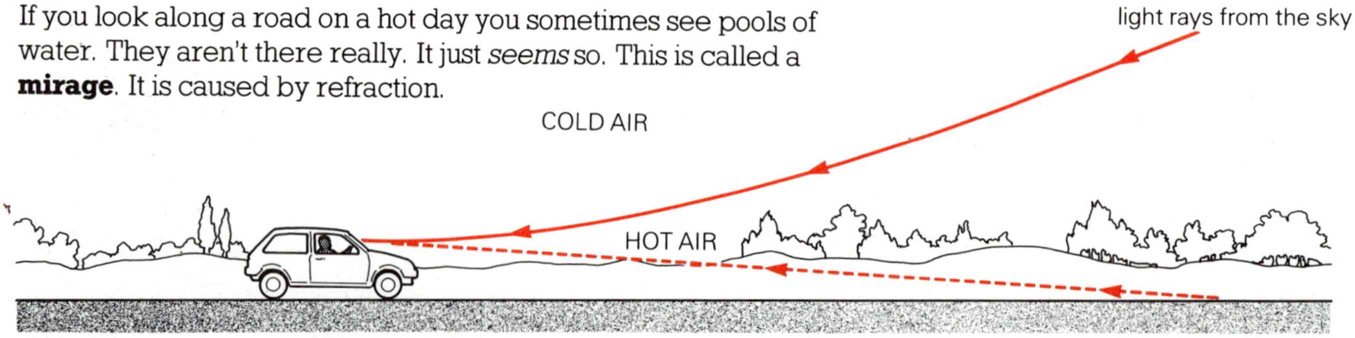

COLD AIR

HOT AIR

light rays from the sky

The light is refracted gradually as it travels from colder air to hotter air. This bends the light rays from the sky into the driver's eyes. He or she sees an image of the sky on the ground. It looks like water in the distance.

26

INVISIBLE MIRRORS

How do they work?

Look at this picture of a ray of light coming *out* of a glass block.

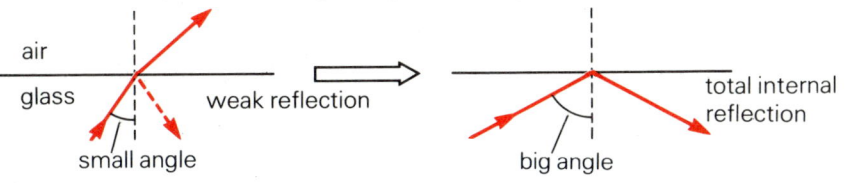

When the angle between the normal line and the surface is too big the ray can't escape. It will be completely reflected back inside. It's as though the side of the glass block was an invisible mirror. This is called **total internal reflection**. It can happen when light is shone through triangular glass blocks called **prisms**. It can be quite useful.

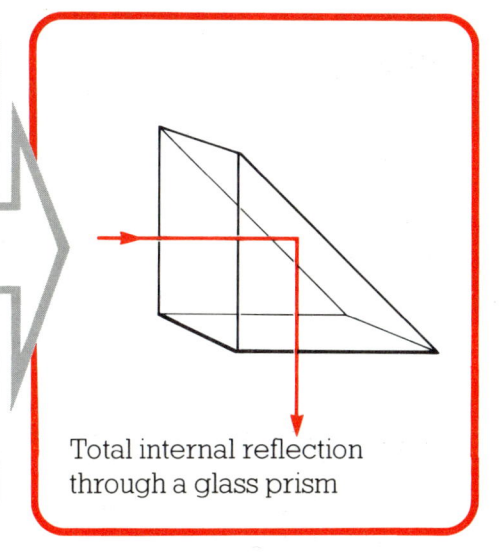

Total internal reflection through a glass prism

Periscopes

You can make a periscope, using prisms that will reflect light by total internal reflection.

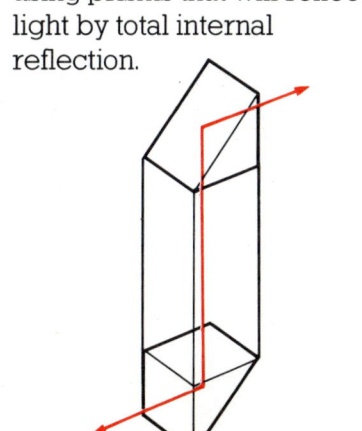

Binoculars

Prisms are used in binoculars. Both the prisms laterally invert the image. So, what you are looking at appears the right way round.

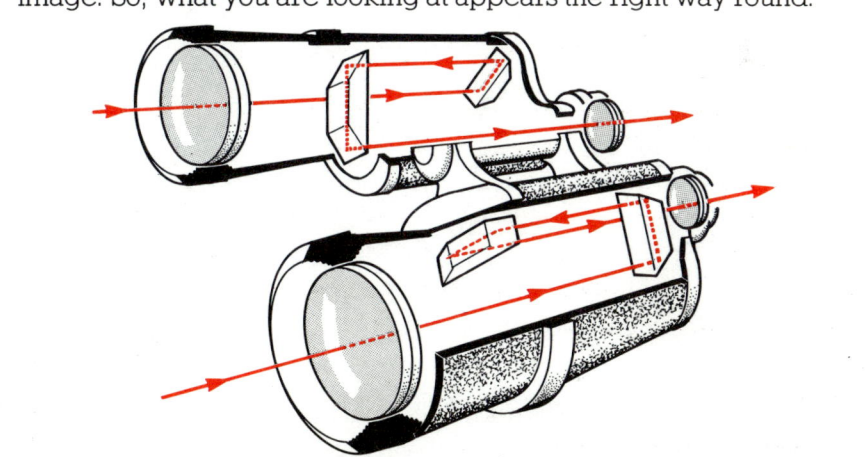

Fibre optics

Optical fibres transmit light over long distances and around corners. They can do this without distorting the image. The light can't escape from the fibre because of total internal reflection.

A bundle of fibres

each fibre carries part of the image

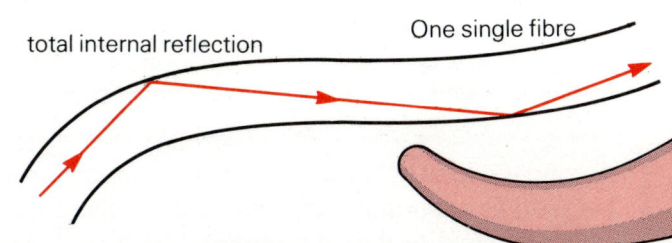

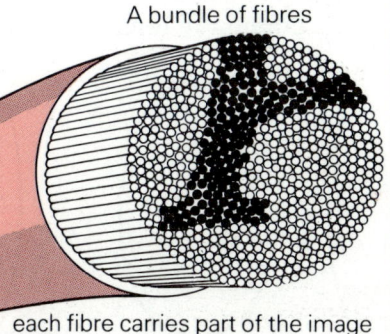

Images are passed along by bundles of fibres. Each fibre carries its own part of the image.

27

LENSES

How do lenses work?

When light is shone through these glass blocks it is refracted. The light rays come to a point. This point is called the **focus**.

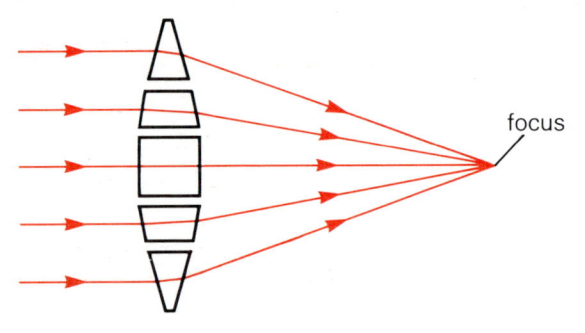

It can also be shone through blocks like these. This time the light appears to have come *from* the focus.

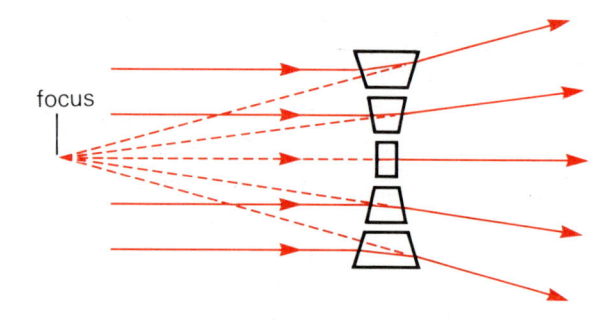

The rays move towards the normal line on the way into the blocks. They move away from the normal line on the way out.

You could join up the small blocks into complete shapes. You would get a **convex lens** and a **concave lens**.

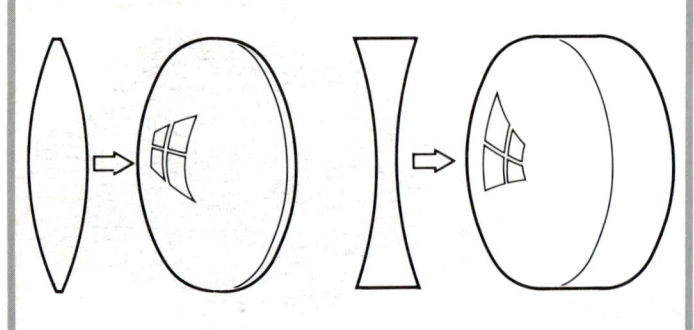

Convex lens   Concave lens

Different images

A convex lens can be used in two different ways. It can be used as a magnifying glass or a projector lens.

The image of the beetle can only be seen *through* the lens. You can't show it on a screen. This kind of image is called a **virtual image**.

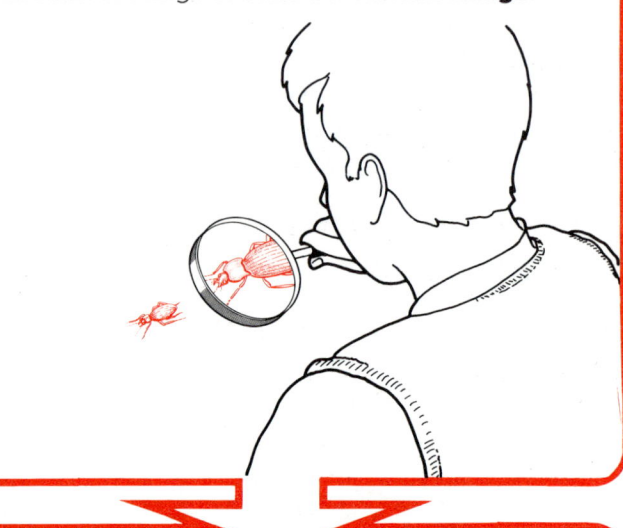

The image of the window can be focused on the screen. It is called a **real image**.

A concave lens only makes virtual images. It can't be used as a magnifying glass because the image is always smaller than the **object** (the thing you're looking at). It won't work as a projector lens either.

Focal length

The distance from the centre of a lens to its focus is called the **focal length**. It's one of the ways we describe lenses. For instance, if you wanted to buy a lens from a supplier you would ask for it by its focal length (amongst other things).

A thin lens has a long focal length. A fat lens has a short focal length.

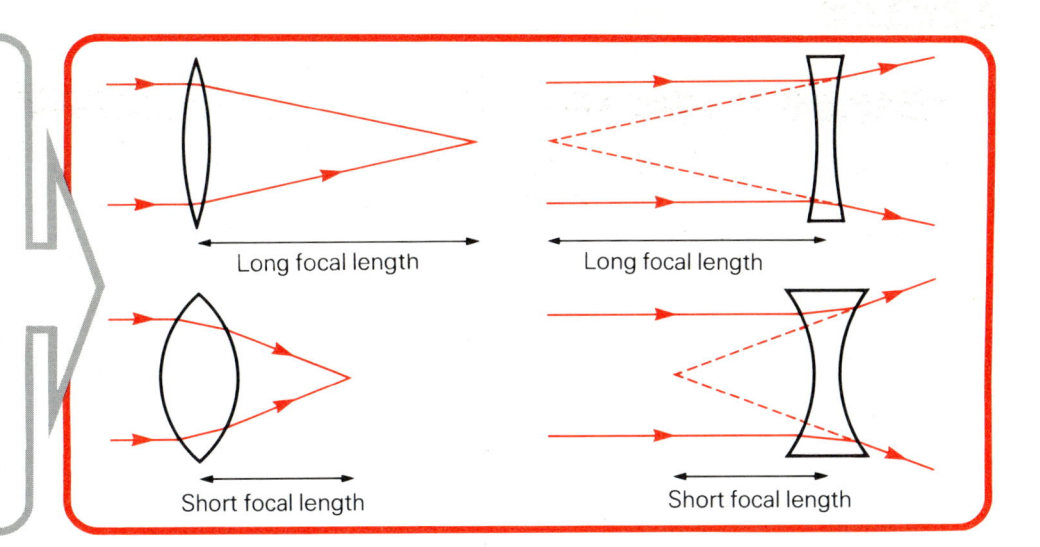

USING LENSES

Lenses are used in many different optical instruments.

Telescopes

You can make a telescope from two convex lenses. One should have a short focal length, the other a long focal length. The long focal length lens must be the one nearest the thing you want to look at. It forms a real image in the middle of the telescope. Instead of focusing this image on a screen, you look at it through the second lens. This magnifies it to give a large final image. Unfortunately this image is upside down!

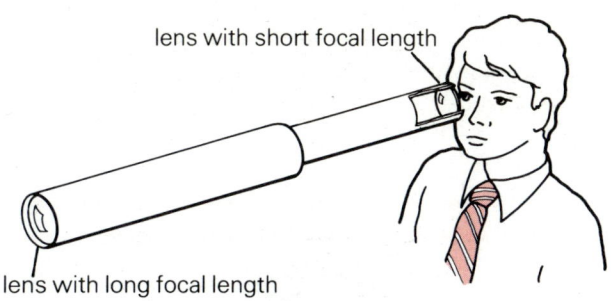

Microscopes

A microscope has two convex lenses just like a telescope. They both have quite short focal lengths though. Which way up is the image in a microscope?

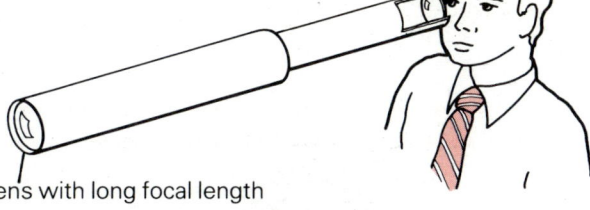

Projectors

A slide projector has a large convex lens. It focuses the image of the slide on the screen. Again, this image is upside down. So, you have to remember to put your slide into the projector upside down when you start!

The convex lens focuses the image on the screen

SEEING THINGS

How your eyes work

Both your eyes contain a remarkable convex lens. It focuses an image on a screen at the back of your eye. Messages about this image are carried by nerves to your brain. Each part of your eye has an important job to do.

Iris
The **iris** controls the amount of light entering your eye.

Cornea
The **cornea** is the clear outer covering at the front of your eye. The cornea surface is curved. So, there is refraction of the light rays here.

Pupil
Light enters your eye through the **pupil**.

Aqueous humor
The **aqueous humor** is a clear liquid in front of the lens.

Focusing-muscles
Muscles are arranged in a ring around the lens. They change the focal length of the lens, by changing its shape. (Remember: a thin lens has a long focal length; a fat lens has a short focal length.)

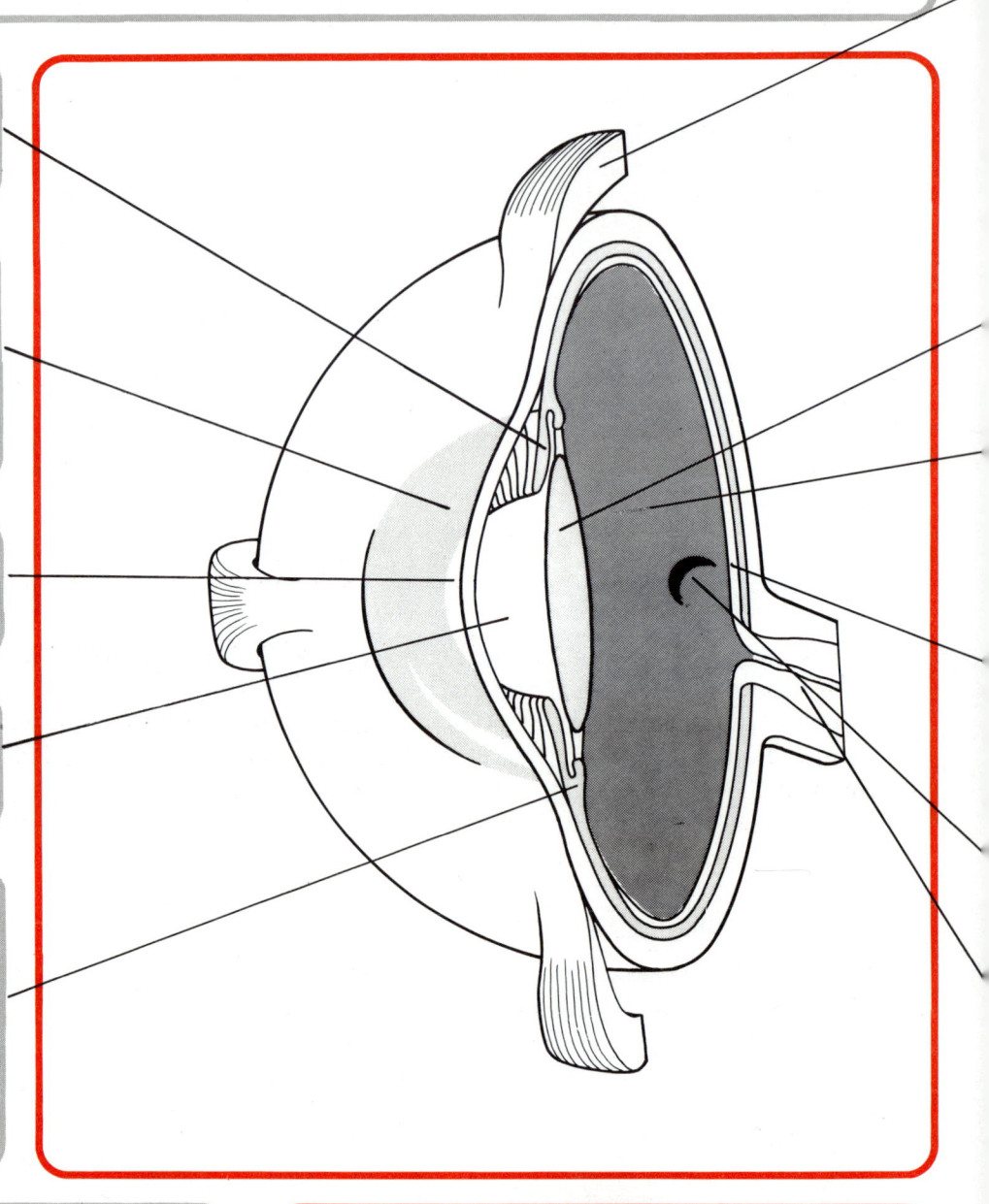

Why is the iris important?

The coloured part at the front of your eye is the iris. The black part in the middle of the iris is the pupil. This is where light rays enter your eye. Your pupil looks black because your retina does not reflect any light back out of your eye. You open up and close down your iris automatically. This controls the amount of light entering through the pupil.

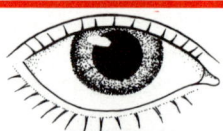

On a dull day the pupil is *large* to let enough light in.

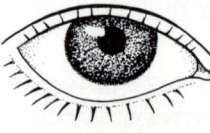

On a bright day the pupil is *small* to stop too much light entering the eye.

30

Eye-moving muscles
Eye-moving muscles turn your eyes towards what you want to see.

Lens
The **lens** focuses the image onto your retina. Its focal length can be changed by the focussing muscles. You can clearly see things that are near you *and* those that are a long way away.

Vitreous humor
The **vitreous humor** is a clear jelly-like substance. It helps to keep the eyeball in shape.

Retina
The **retina** is the screen at the back of your eye. It is where the image is focused. It is made up of millions of **receptors**. These are special light sensitive cells. They are joined to the **optic nerve**.

Fovea
The **fovea** is the most sensitive part of your retina.

Optic nerve
The **optic nerve** sends messages to your brain about the image on your retina.

Focusing the image

If you look at anything a long way away, the lens in your eye will be thin. The muscles will have stretched the lens until the image is focused on your retina.

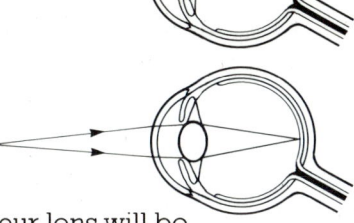

If you look at anything near you, your lens will be fat. The muscles have contracted and 'bunched-up'. This makes the lens fat enough to focus the image on your retina.

CAMERAS
How do they work?

The camera has a convex lens. It focuses the image onto the film. The focal length of this lens is fixed. The lens has to be moved backwards and forwards to change the focus.

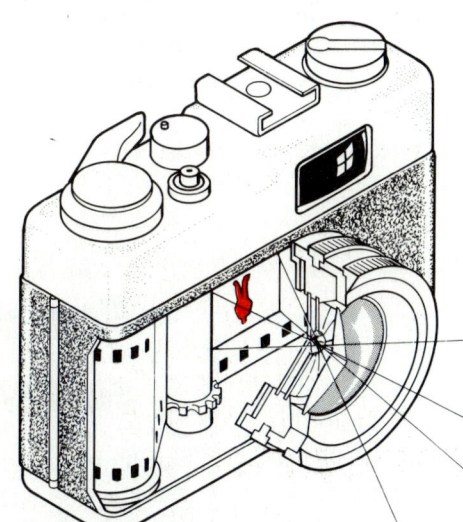

When you take a picture – of your friend, maybe – the **shutter** opens briefly. This allows light from your friend to reach the film. Chemicals on the surface of the film react to light. The film must now be treated with other chemicals before you can see the picture on it.

The **aperture** can be adjusted so that the film is exposed to the right amount of light. It's just like the pupil in your eye. The aperture has to be small on a bright day and large on a dull day.

31

USING A PIN-HOLE CAMERA

In this Activity you are going to make a pin-hole camera. It will give you more evidence to show that light rays travel in straight lines. When you have made the camera, you can try several things to find out how it works. You may also be able to take photographs with it.

You will need
- two cardboard tubes or boxes
- aluminium foil
- tracing paper
- scissors
- convex lens
- light bulb
- sticky tape
- pin

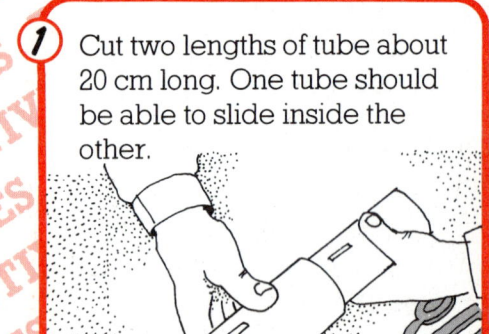

1 Cut two lengths of tube about 20 cm long. One tube should be able to slide inside the other.

2 Separate the tubes. Stick tracing paper over one end of the smaller tube. Stick aluminium foil over one end of the larger tube. Make a pin hole in the middle of the foil.

3 Slide the smaller tube inside again. Point the camera towards something that is well lit and interesting. Look through the small tube. Describe what you can see.

Things to try

1 Twist the tubes in opposite directions. Does the picture stay the same?
2 Point the camera towards a light bulb with a large filament. Make two pin holes in the aluminium foil. How many images can you see? Try three pin holes. How many images are there now?
3 Make the pin hole larger by pushing a pencil through the foil. What happens to the image?
4 Put a convex lens in front of the pencil hole. Move the lens backwards and forwards. What happens to the image?

Taking photos

You can take real photos with a pin-hole camera. But it must be light-tight! If light can get into the camera you have made, you may have to use a tin instead of the tubes. In a darkroom, replace the tracing paper with photographic paper. Cover the pin hole. Point the camera at an interesting scene. Expose the paper to light by uncovering the pin hole for a minute or so. Then, develop the photographic paper in the darkroom, using developing solution and fixer. You will have to experiment with exposure and developing to get the best picture.

How does the pin-hole camera work?

The pin-hole camera works because light from the scene reaches the camera by moving in straight lines. The rays cross over at the pin hole. Turning the camera upside-down makes no difference to the image!

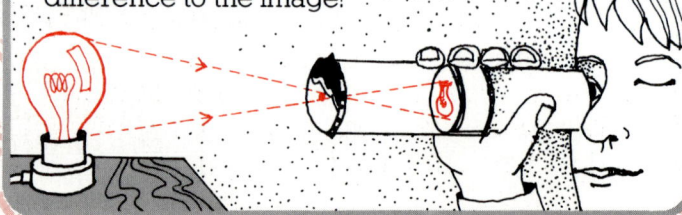

FINDING OUT ABOUT REFLECTION

Light always obeys the same rules when it reflects from a surface. In this Activity you are going to find out about these rules of light reflection.

You will need
- ray box and slit
- white paper ● pencil
- ruler ● mirror
- protractor ● plasticine

Copy this table

Angle of incoming ray/degrees	Angle of reflected ray/degrees

1. Draw a line across the white paper. Draw another line at right angles to the first one. The first line marks the mirror position. The second line is the normal line.

2. Stand the mirror upright on the paper. Have the back of the mirror running along the first line.

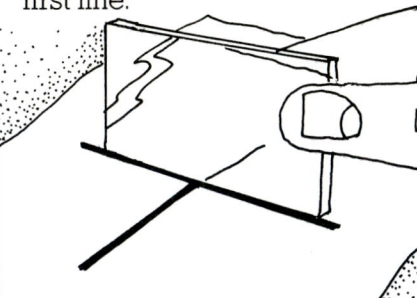

3. Shine a ray of light from the ray box onto the mirror. Aim it at the point where the normal line meets the mirror.

4. Mark the path of the ray, and its reflection, with small crosses. Remove the ray box and mirror.
   Draw the path of the light rays through the crosses.

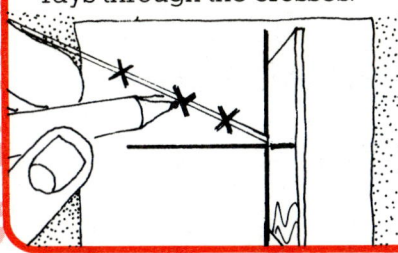

5. Measure the angle between the incoming ray and the normal line. Write the angle in the table. Repeat this with the reflected ray.

6. Change the angle of the incoming ray. Repeat steps 3 to 6.

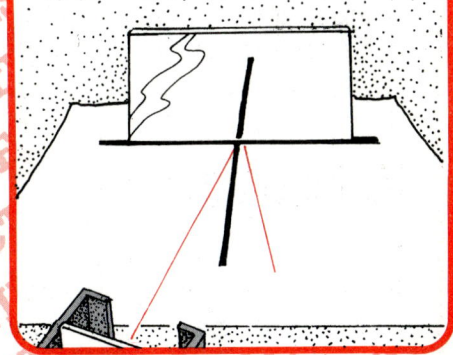

Answer these questions

1. Look at each pair of results. What do you notice about the incoming ray and the reflected ray?
2. Why did you need to put the back of the mirror on the line?
3. What do you think would happen if you shone the incoming ray along the normal line? Try it.

USING MIRRORS

Mirrors are very useful and good fun too! Here are some things to try.

You will need
- cardboard
- scissors
- two mirrors
- sticky tape
- glue
- small coloured things

Making a periscope

1. Copy this shape onto a large piece of cardboard. The width of each strip should be the same as the width of your mirrors.

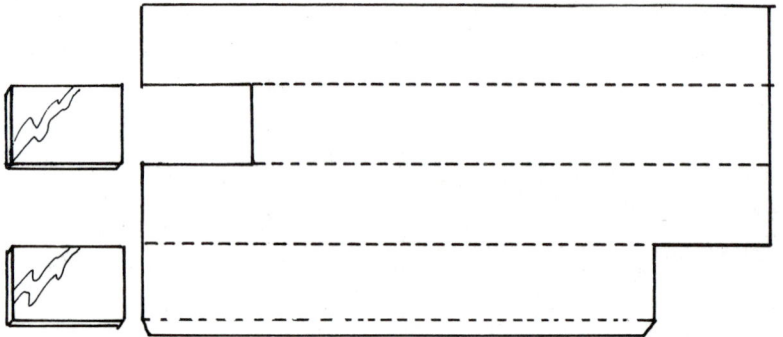

2. Fold the cardboard to make a long thin box. Glue it together.

Fix the mirrors inside the tube with sticky tape. They must both be at 45° to the side of the tube.

Answer these questions

1. Which way up is the image you see using the periscope?
2. What would you use the periscope for? When would you use it?
3. Explain how the periscope works.

Making a kaleidoscope

Put the mirrors facing one another, with angle of 60° in between. Sprinkle a selection of coloured things in the space in between. Look at them from above.

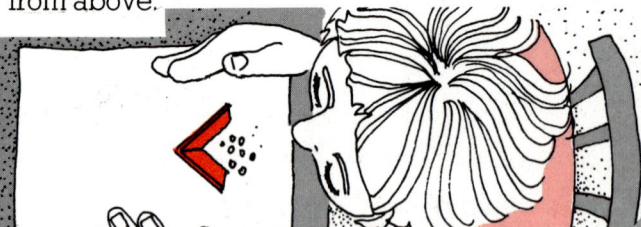

Answer these questions

4. How many times can you see one of the objects being reflected?
5. How could you improve your kaleidoscope?

See yourself as you are!

When you look in a mirror, you usually see a mirror image of yourself. Try this and you'll see yourself as you *really* are. Hold two mirrors with an angle of 90° between them. Look where the mirrors meet.

Answer these questions

6. How does this work?
7. Are photos of you the same as the image other people have of you?

FINDING OUT ABOUT REFRACTION

Many substances let light pass through. As a light beam passes from one of these substances to another it may change direction. This is called **refraction**. This Activity is an investigation into refraction.

You will need

- rectangular glass block
- semi-circular glass block
- ray box and slit
- white paper • pencil
- ruler • protractor

1 Put the rectangular glass block in the middle of the paper. Draw round it. Shine a ray through the block like this.

2 Mark the path of the ray with crosses. Remove the ray box and the block. Draw the path of the ray. Add normal lines at the places where the ray entered and left the block.

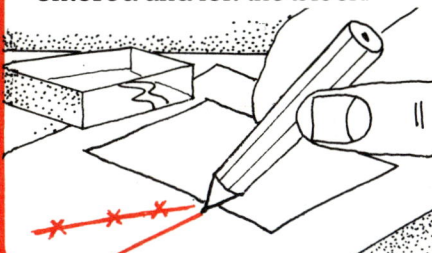

3 Change the angle of the light ray. Repeat steps 1 and 2.
Try to find a rule about how light changes direction when it enters and leaves the block.

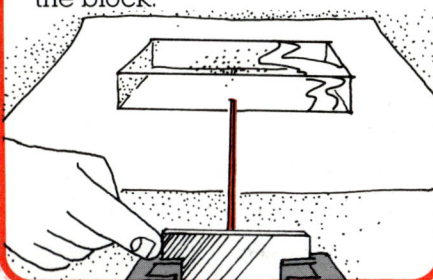

4 Put the semi-circular block on a fresh piece of paper. Draw round it. Shine a ray in through the curved side towards the centre of the flat side.

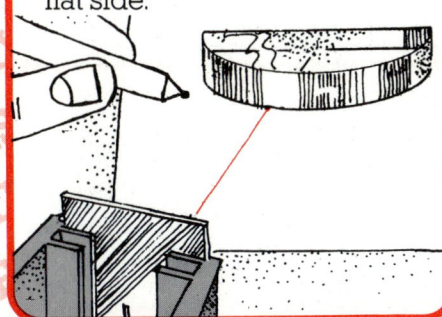

5 Change the place where the ray enters the glass block. Make sure you keep aiming at the centre of the flat side.
Move the ray until you just reach the point where no light escapes from the flat side. Mark out the paths of the light rays when this happens. Draw a normal line on the flat side. Measure the angle between the incoming ray and the normal line. Label this angle C.

Answer these questions

1. Can you think of any rules about the change of direction of the light ray when it entered and left the rectangular glass block?
2. Angle C is called the **critical angle**. What size was it?
3. Why didn't the light change direction when it *entered* the semi-circular glass block?

The critical angle

At the critical angle, the angle between the refracted ray and the normal line is 90°. If the angle of the incoming ray is more than the critical angle, you can't see any light escaping from the flat side. It is all reflected back inside the block. This is called **total internal reflection**.

FINDING OUT ABOUT LENSES

Different lenses give different images. The sort of image you get depends on the type of lens and where you place the thing you are looking at.

You will need
- convex and concave lenses
- lens holder
- metre ruler

Convex lenses

Copy this table

What did the lens look like?	Focal length /cm	Size of image/cm	Which way up was the image?	Was it a good magnifying glass?

1. Look at the lens. Is it fat or thin? Describe it in the table.

2. Hold the lens next to a wall opposite a window. Move the lens slowly away from the wall until you can see a clear image of the window on the wall.

3. Measure the size of the image of the window. Decide which way up it is. Write your results in the table.

4. Measure the distance between the lens and the wall. This is almost equal to the focal length of the lens.

5. Use the lens as a magnifying glass. Decide whether the lens is a good magnifying glass or not. Write what you decided in the table.

6. Repeat steps 1 to 5 with other lenses. Answer questions 1 and 2.

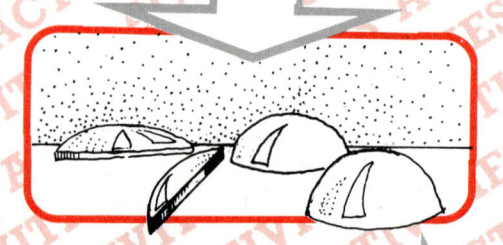

Go on to the next page ➡

36

Carry on here →

Answer these questions
1. Make a summary of your results.
2. You have seen two different types of image. One you could focus on the wall, the other you could only see by looking through the lens. What are the names of these two types of image? (Look at page 28.)

Concave lenses

1. Try to focus the image of a window on a wall using a concave lens. Follow steps 1 and 2 in the Activity on page 36. Describe what you see.

2. Try using the lens as a magnifying glass. Describe what you see. Answer questions 3 and 4.

Making a telescope

Hold a convex lens, with a long focal length, at arm's length. Look at it to see the first image. It should be upside down. Use a second convex lens, with a short focal length, as a magnifying glass to look at the first image. Answer question 5.

Replace the second lens with a concave lens. What can you see through this telescope?

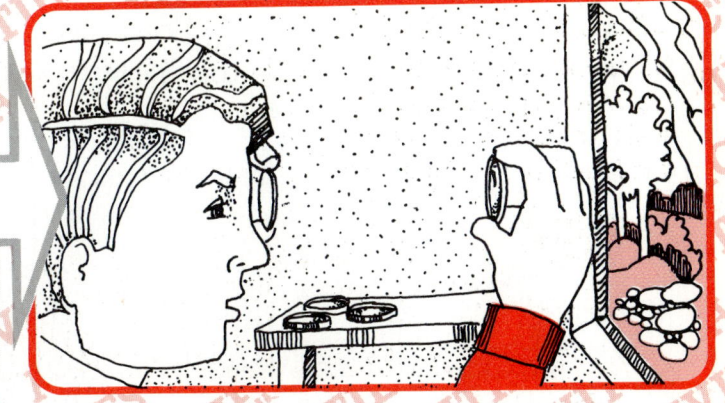

Answer these questions
3. Was the concave lens a good magnifying glass?
4. What sort of images do you get with a concave lens?
5. Describe the image you saw through the first telescope.

EXERCISES

1 Draw a diagram that shows an eclipse of the Moon by the Earth.

2 (a) Which capital letters of the alphabet look the same in a mirror?
(b) Suppose you wanted to put 'ACME OVERTROUSER COMPANY' on the front of a van so that it could be read correctly in a driving mirror. What would it look like?

3 Why does Steve Davis need to know the law of reflection?

4 What is angle *x*? You can redraw this accurately or work out an answer.

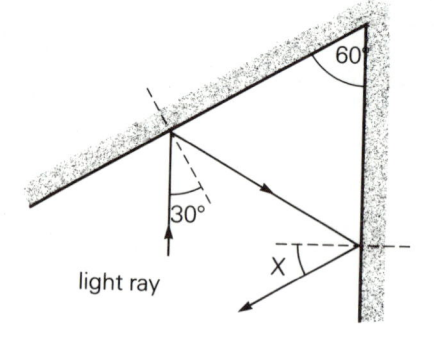

5 (a) Copy this diagram of a pin-hole camera to scale. Find the height of the image of the man.

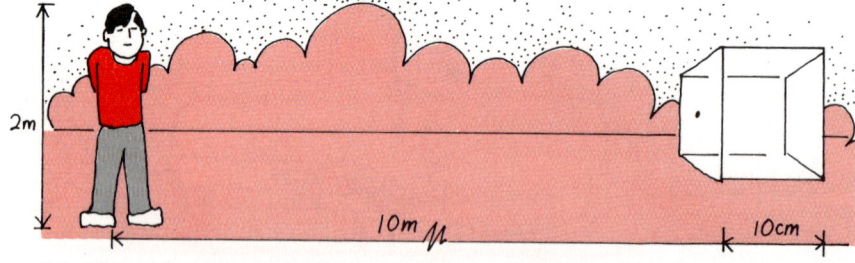

(b) What happens to the image in a pin-hole camera as you walk towards the object?
(c) How does the image change when you move further away? Draw ray diagrams to explain your answer. If you turned the pin-hole camera upside down what effect would it have on the image?

6 How would you measure the focal length of a convex lens? Look at these lenses. Which has the largest focal length?

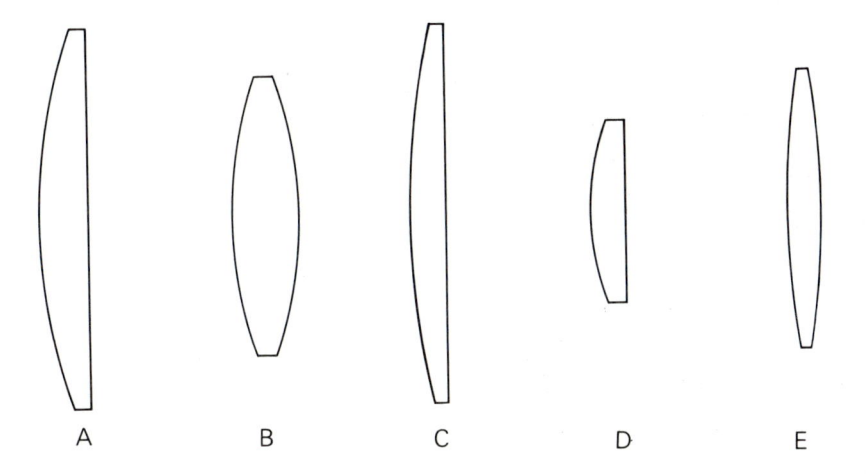

A B C D E

7 Pair up these words. Explain why you have paired them.

Eclipse
Optical fibre
Convex mirror
Mirage
Slide projector
Concave lens

Refraction
Driving mirror
Shadow
Total internal reflection
Virtual image
Convex lens

8 Fill in the missing labels. Explain what each part you have labelled does.

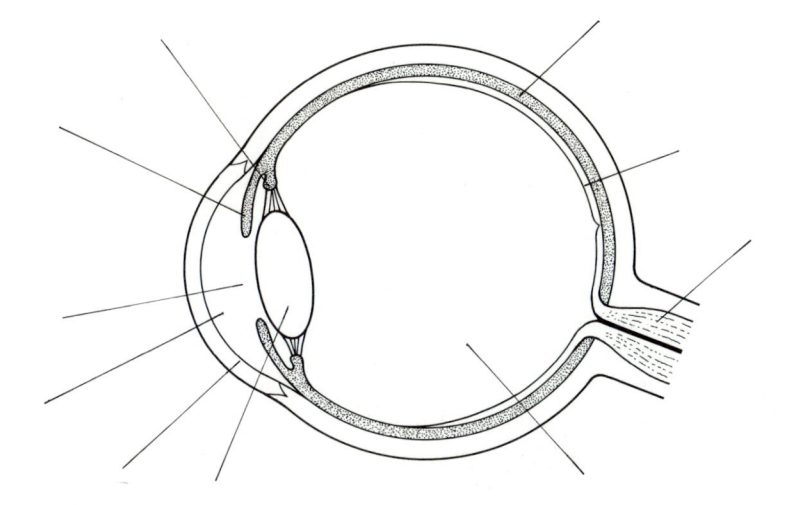

9 Fill in the missing words.

Refraction
When a beam of ……… enters a rectangular glass ……… it bends ……… the normal line. As it leaves, it bends ……… the ……….

3 WAVES

INTRODUCING WAVES

Most people think of the seaside if you ask them about waves. But there are many different kinds of waves apart from those in water. For example, sound, sunlight and radio signals are all waves. Water waves do have a lot in common with the other waves, though. A close look at water waves will tell you a lot about the others.

MEASURING WATER WAVES

The girl in the picture has thrown a stone into the pond. The stone had energy as it fell into the water. Some of that energy is now being moved outwards by the circular waves. All waves do this. They all move energy from one place to another.

As the waves go past, the boat will bob up and down. The same thing happens to the water as the wave moves along. Each part of it moves up and down in turn. This passes the wave through the water.

A closer look at water waves

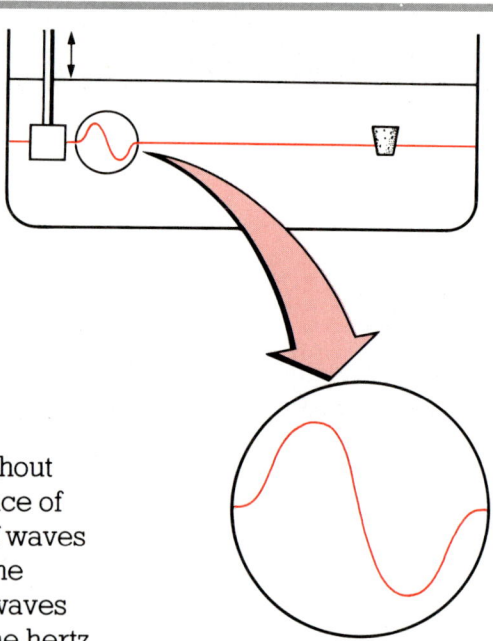

Look at this picture.

It's a side view of a tank of water with oil on top. Imagine that you moved the plunger down into the water, back up into the oil and then back to where it started. This would send a single wave along the surface of the water. The movement of a single wave is called a **cycle**.

The wave would soon reach the cork. The cork, in turn, would copy the movement of the plunger.

Now imagine that you moved the plunger up and down without stopping. This would send *many* single waves along the surface of the water. It would make a **continuous wave**. The number of waves you made in a second is called the **frequency** of the wave. The faster you moved the plunger, the higher the frequency. All waves have a frequency. It's measured in **hertz**. (You say 'hurts'.) One hertz (Hz) is also called one **cycle per second**.

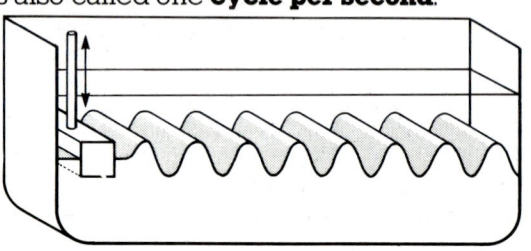

To measure the frequency of water waves you need to count the number of single waves that go past in a second.

40

MEASURING WATER WAVES

Water waves come in all shapes and sizes! You could measure lots of different things about them. But there are three things it's very useful to know. These are:
- the size of the wave,
- how fast it is going,
- the length of each wave.

How big?

The size of a wave is called its **amplitude**. It is the largest movement made by the water from the flat surface. So, sea waves have a bigger amplitude than the ripples on a pond.

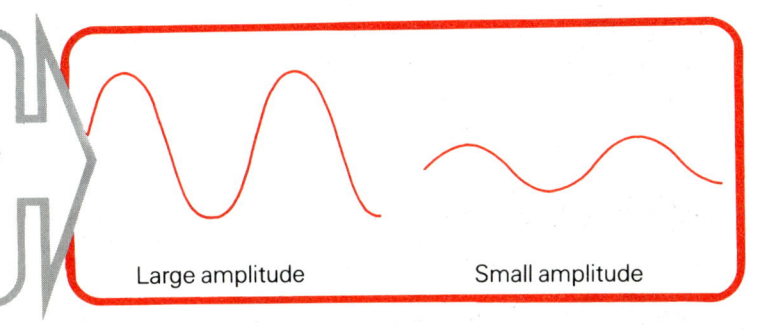

Large amplitude   Small amplitude

How fast?

To find out how fast a wave is moving you need to know how far it has moved in a certain amount of time. You then divide

$$\frac{\text{distance moved}}{\text{time taken}}$$

This tells you how fast the wave is going. We call this its **velocity**. Look at this picture. It shows two people measuring the velocity of waves at the seaside.

The boy signals as a wave passes him. The girl measures how long it takes to move 50 m. What is the wave's velocity?

How long?

We call the length of a wave the **wavelength**. It is the distance from any place on a wave to the same place on the next wave.

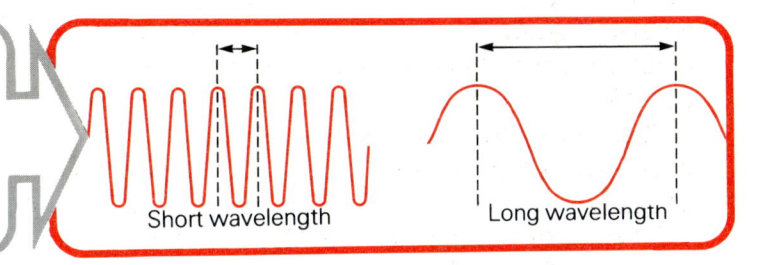

Short wavelength   Long wavelength

It's hard to measure the wavelength of waves with a ruler. They are moving all the time. There is an easier way using the velocity and frequency. The frequency multiplied by the wavelength is the velocity of the waves:

$$\text{Wavelength} = \frac{\text{velocity}}{\text{frequency}}$$

This equation is known as the **wave equation**.

Transverse waves

We call waves in water **transverse waves**. This is because the water moves at right angles to the direction in which the wave is moving.

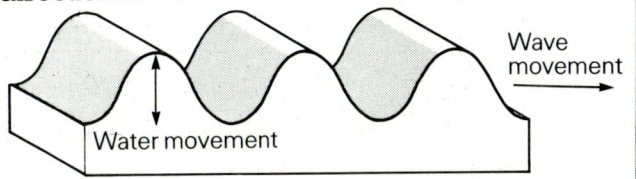

Water movement   Wave movement

41

SOUND

What is sound?

The world around us is a very noisy place!

How many of these sounds have you heard? They are all very different. What makes them different? How do these sounds reach you? To find out the answers to some of these questions, we need to find out what sound is and how it travels.

Sound waves

Sounds travel from place to place as **sound waves**. Usually the waves are passed on by vibrations of the air.

Sound waves are not transverse waves like the ripples on the pond. They are **longitudinal waves**. This means that the air vibrates backwards and forwards *in the same direction* as the wave is moving. It doesn't vibrate up and down like the water in the pond.

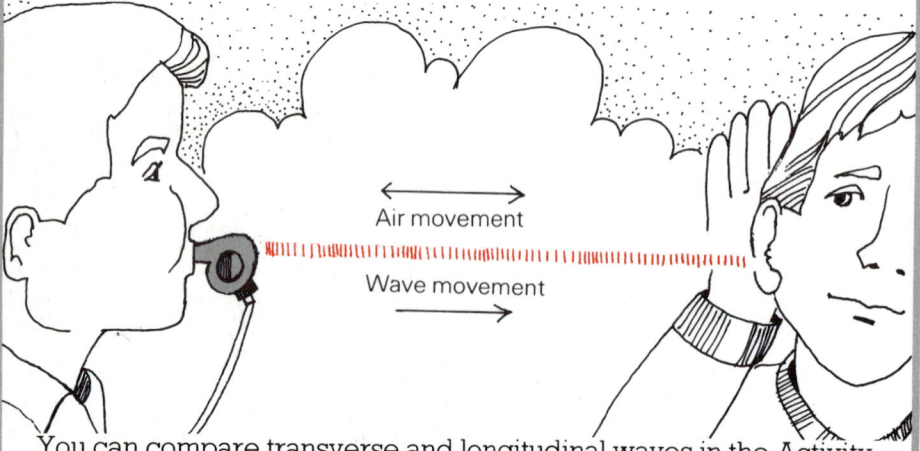

You can compare transverse and longitudinal waves in the Activity on page 52.

Sound travelling

Sound must have a substance to travel through. It cannot pass through a vacuum. There's nothing in a vacuum to pass on the vibrations. You can show this easily.

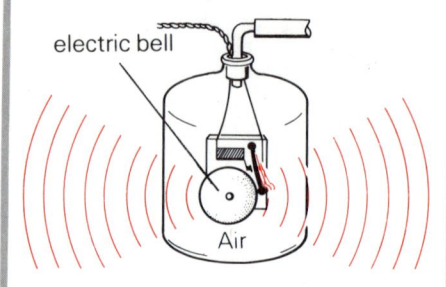

The ringing of the bell escapes through the air and glass.

But when the air is pumped out . . . no sound is heard.

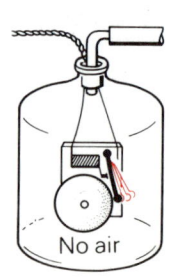

This is why you can't hear the roar of the Sun!

Sound travels better through substances that are thicker than air. You can listen to the workings of an engine or motor using a 'screwdriver stethoscope'.

42

ECHOES ECHOES ECHOES ECHOES

Echoes

Echoes are the reflections of sound waves. They can be useful, but sometimes they are a nuisance. For instance, echoes in concert halls and recording studios can spoil the music. These places have to be carefully designed so there aren't too many echoes.

This picture shows a room that has no echoes. It is called an **anechoic room**. The ridges on the walls stop the reflection of the sound waves.

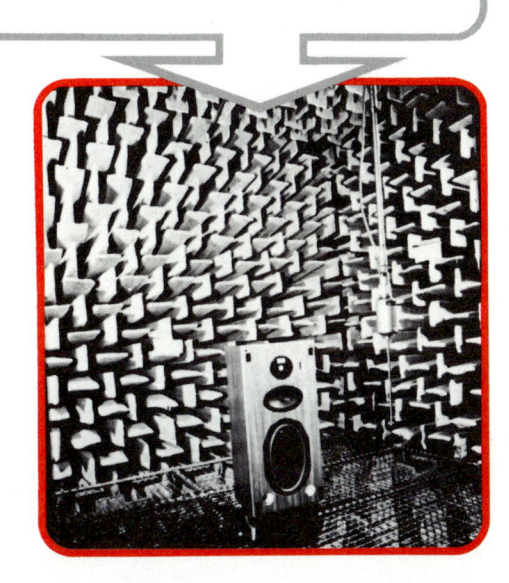

Stationary waves

You can make stationary longitudinal waves with a 'slinky' spring fixed at one end. If you move your hand backwards and forwards the waves you make match the reflected waves. It looks as though the waves aren't moving. They seem to be stationary. For this to happen you've got to move your hand at just the right frequency.

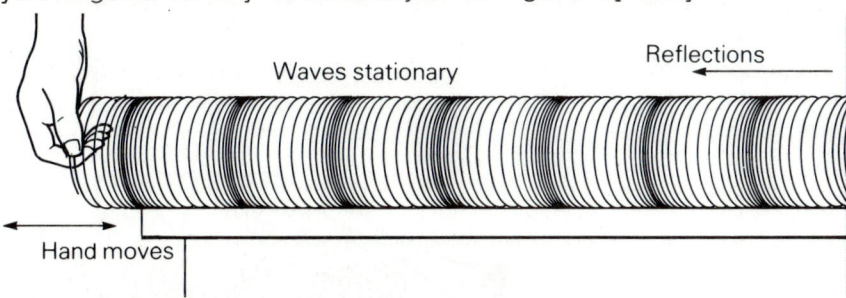

You can try this in the Activity on page 52.

You can make stationary waves by blowing across the top of an empty bottle. If you blow at the right angle you will hear a sound. These are stationary sound waves in the bottle. The waves you make match the waves reflected from the bottom of the bottle. The air in the bottle is **resonating** at its natural frequency.

Echo sounders

Echo sounders on ships are used to find things underwater. They work like this.

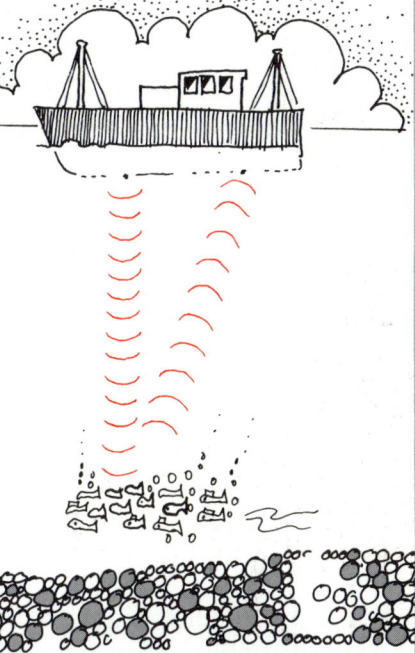

The ship has a sound wave transmitter and a receiver on the hull. The sound waves are sent out through the water.

They may reflect back from the sea bottom or a shoal of fish or an undiscovered wreck.

Sound waves travel through water easily. This is why echo sounders work so well.

43

MORE ABOUT SOUND

Pitch and frequency

The violinist is playing a note called middle C. The string on her violin is vibrating backwards and forwards 256 times a second. This makes a sound wave with a frequency of 256 Hz. If you could hear her now, the air just outside your ear would be vibrating at 256 times every second.

The sound of a musical note is known as the **pitch**. The higher the frequency, the higher the pitch.

Octaves

This singer is reaching top C. This note is one **octave** higher than middle C. Her vocal cords are making the air molecules vibrate at 512 Hz. This is *twice* the frequency of middle C.

If the frequency of a sound wave is doubled, then the pitch goes up by one octave.

Tuning forks

A tuning fork is made to make a note of a certain pitch. The frequency of the note it makes is stamped on it just below the prongs.

Which note will this tuning fork make?

Seeing sounds

Microphones change sound waves into waves of electricity. These waves can then be seen as transverse waves on a special TV, called a **cathode ray oscilloscope**. In this way, it's actually possible to 'see' the sound waves from a musical instrument.

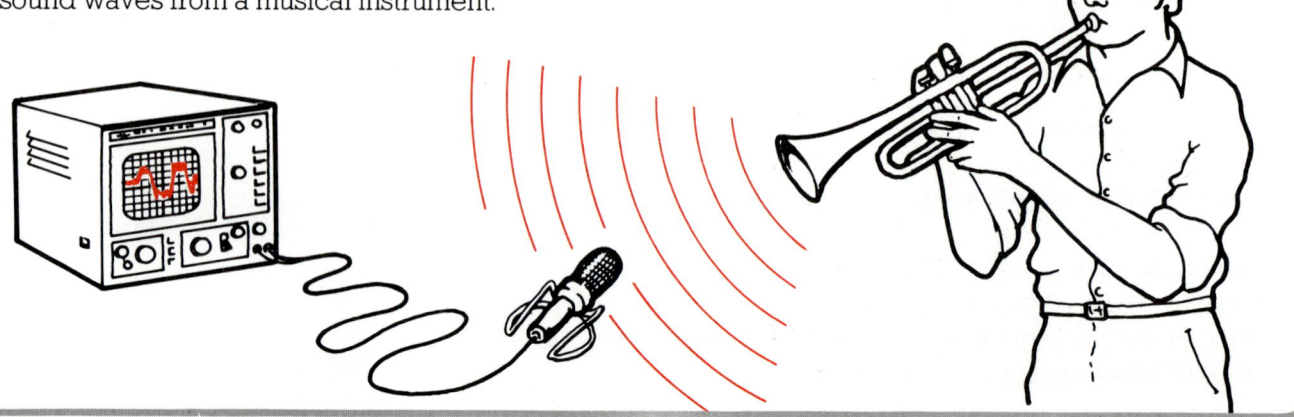

Loudness and amplitude

You can use a microphone and an oscilloscope to see the differences between notes. Look at these two notes. The amplitudes of the waves are very different.

The loudness of a note depends on the amplitude of the sound waves; the louder the note, the bigger the amplitude. Loud notes also contain more energy than quiet ones.

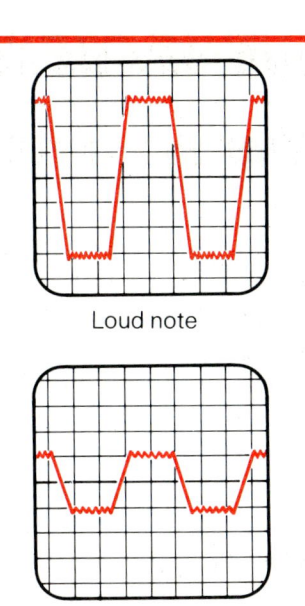

Loud note

Quiet note

Different sounds

The same notes played on different musical instruments sound different. Think of a clarinet and a violin playing the same tune. Middle C played on a clarinet looks like this.

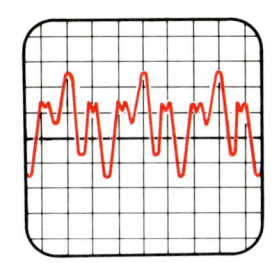

But the same note on a violin looks like this.

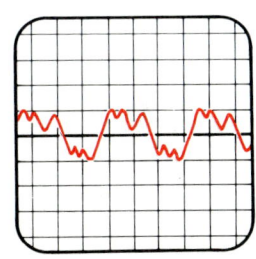

Compare the waves from the clarinet and the violin with middle C from the tuning fork.

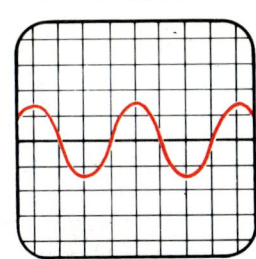

All these waves have the same basic frequency – 256 Hz. But the clarinet and violin have had other waves added.

It is these additions to the basic notes that give instruments their different sounds. These additions are called **harmonics**. The quality of the note from a musical instrument depends on the harmonics.

Synthesizers

Sound synthesizers make different waves electronically. Some even use computers to control how they work.

45

HEARING SOUNDS

How do you hear?

Think of all the millions of different sounds you can hear. How can you tell them apart? You use your ears, of course, but they are just the start of the process. Inside your head is some amazing 'machinery' that changes sound waves into nerve messages. Your brain then sorts these messages out.

This part is made of skin and cartilage. In some animals it directs sound waves towards the ear-drum. In humans, though, it's not very important.

These three bones pass on the ear-drum movements to the oval window.

This nerve carries the nerve messages from the cells in the cochlea to your brain.

The **cochlea** is the part of the inner ear that changes vibrations into nerve messages. It is a long tube wound up into a spiral. It contains cells which are sensitive to vibrations of the liquid in the tube.

Your **ear-drum** vibrates when sound waves hit it.

This tube connects your middle ear to the back of your throat. It makes sure the air pressure inside your middle ear is the same as the air pressure outside.

The **oval window** is a thick skin which separates the air-filled middle ear from the liquid-filled inner ear. Movements of the oval window pass on the vibrations into the liquid in the inner ear.

Deafness

Any break in the 'chain' between the ear-drum and the brain causes deafness. If wax blocks the ear tube, sound waves can't reach the ear-drum. A loud explosion may split the ear-drum. Sometimes, the tiny bones in the middle ear seize up. A hearing aid or an operation can help.

Loud noises of about the same pitch can damage one small part of the cochlea. Some pop musicians are said to suffer from this. A few illnesses can damage the cells in the cochlea or the nerves that connect them to the brain. Hearing aids can't help when this happens.

This man is wearing ear protectors to stop the noise damaging his ears

How do you speak?

The sounds your voice makes come from your **vocal cords**. These are thin strips stretched across your voice box (larynx). As you blow air past them, they vibrate. This makes a sound. To make high notes you automatically stretch your vocal cords tighter. To make low notes you slacken them. The length of the cords also affects the pitch. The frequency of the vibration is between about 100 and 500 Hz. Men usually have deeper voices than women. This is because their voice boxes – and their vocal cords – are usually larger.

Your tongue, lips and teeth work together to form the different sounds when you speak. How are they involved when you say the words 'but', 'tip', 'did'? What do they do when you say 'air' and 'ear'?

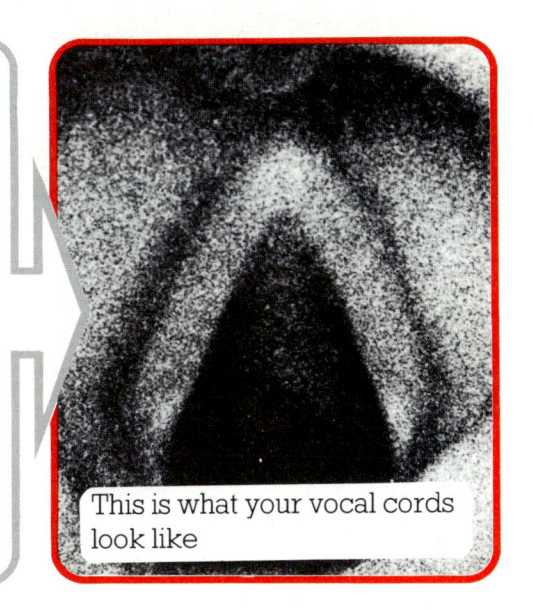

This is what your vocal cords look like

Bats

Bats are active at night. They are nocturnal. At night they can't see well enough to avoid flying into things. They can't see their prey either. So, they use their voices. They make sounds with frequencies between 60 000 and 200 000 Hz. These sounds reflect from different objects. Their ears pick up the reflections.

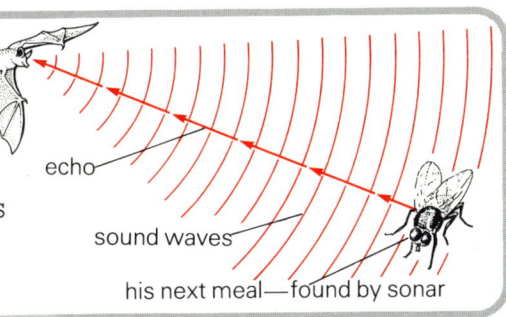

echo
sound waves
his next meal—found by sonar

The Doppler Effect

Have you ever noticed what happens to the sound of an ambulance siren as the ambulance drives past? The sound from the siren changes. The pitch is higher as the ambulance approaches. It then drops to a lower pitch as the ambulance passes you and goes away. This is called the **Doppler Effect**. It's easy to understand how this happens when you know about sound waves! As the ambulance moves towards you it 'squashes up' the waves. As it goes away, it 'stretches them out'.   So, the note *sounds* higher as the ambulance comes towards you. It then *sounds* lower as the ambulance goes away.

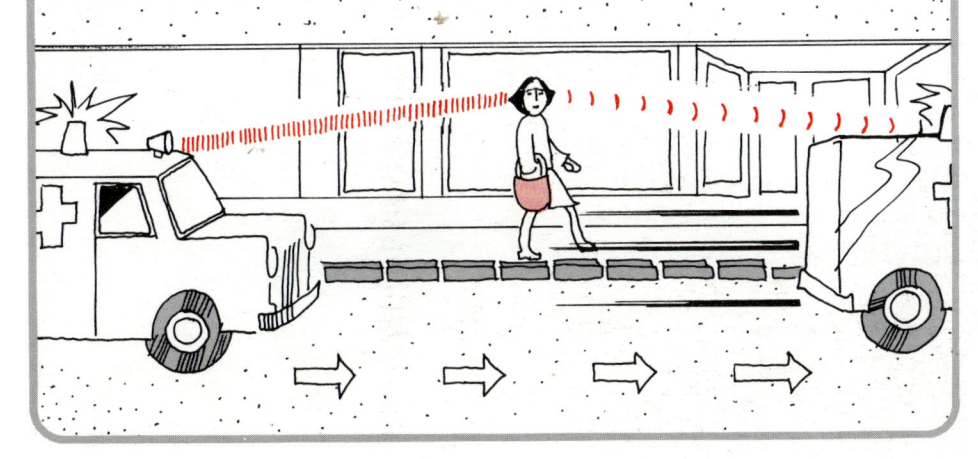

USING A RIPPLE TANK

You can find out a lot about waves by looking at ponds or going to the seaside. To find out some things about waves though, it's easier to use a 'portable pond' you can bring indoors! We call it a **ripple tank**. It gives you a very good view of waves and what they do.

You will need

- ripple tank
- darkened room
- wooden rod
- lamp • pipette
- motor fixed to a beam
- support rods • dipper
- hand stroboscope
- power supply
- elastic bands

1. Fit the legs onto the tank. Pour water in to a depth of 1 cm. Place the lamp so that it shines on the white surface underneath.

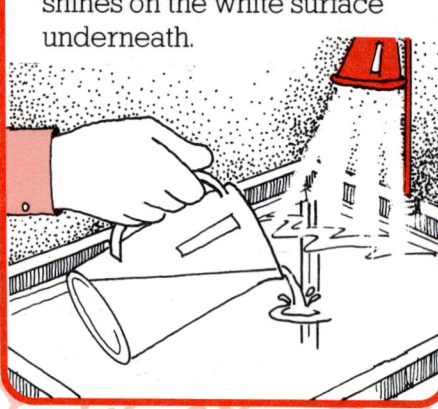

2. Use a dropping pipette to make circular waves.

3. Roll the wooden rod backwards and forwards in the water. This will make plane waves.

4. Fix the motor supports and dipper together so that the dipper lightly touches the water.

5. Connect up the motor to the power supply. Set the power supply to the correct voltage. (Ask your teacher for help). Switch on.

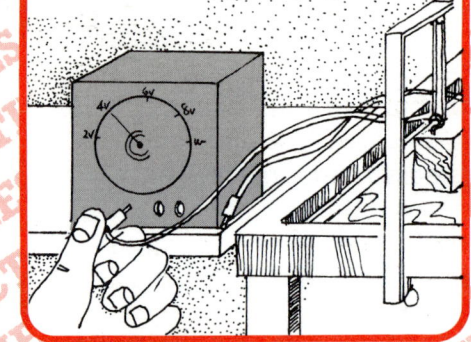

6. Use the hand stroboscope to 'freeze' the action. Look through the stroboscope at the white surface. Turn the stroboscope until the pattern you see appears to stand still.

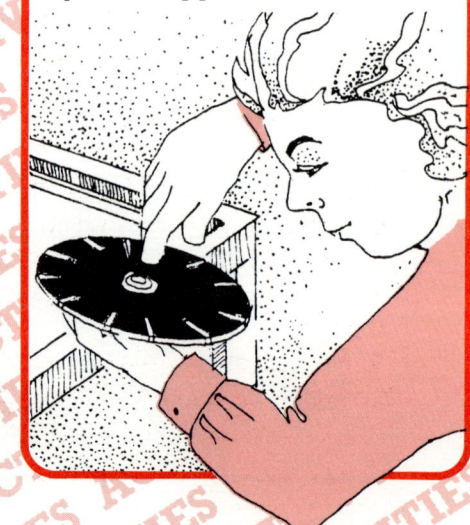

Answer these questions

1. Explain how the waves show up on the white surface.
2. What moves from the drop of water to the edges of the ripple tank when you make circular waves move through the water?
3. How can you change the wavelength of the water waves? How would a change show up on the white surface? (You can try this with the plane waves.)
4. How can you change the amplitude of the water waves? (Test your answer using circular waves.)
5. How does the stroboscope 'freeze the action'?

REFLECTIONS IN A RIPPLE TANK

This Activity follows on from the Activity on page 48. It shows you how you can find out what waves do when they bounce (reflect) off barriers in the water.

You will need
- ripple tank
- darkened room
- lamp
- motor fixed to a beam
- support rods
- dipper
- hand stroboscope
- power supply
- flat and curved barriers
- elastic bands

1. Arrange the motor beam so that it just dips into the water along all of its length. Switch on the motor.

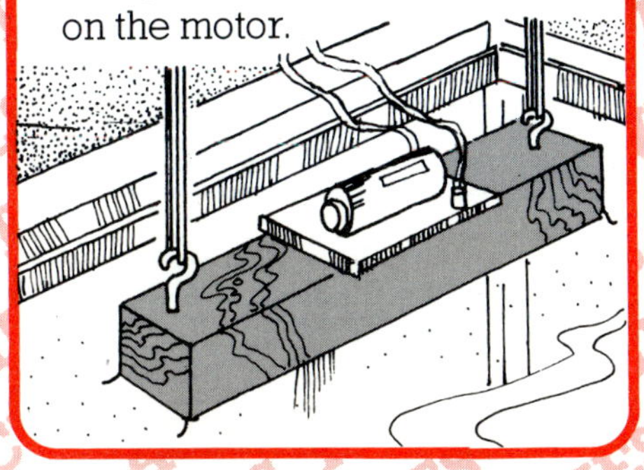

2. Put the flat barrier in the ripple tank so that the waves reflect off it.

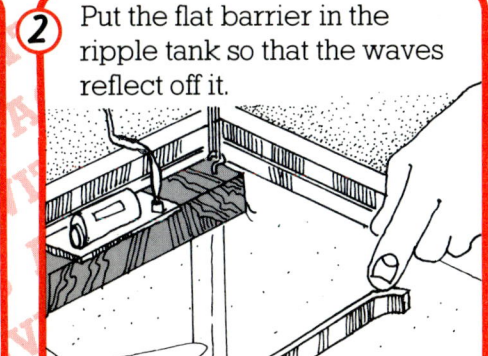

3. Change the angle between the flat barrier and waves to about 45°. 'Freeze the action' with the stroboscope.

4. Send plane waves towards a curved barrier. Use the stroboscope to see what happens.

5. Turn the curved barrier around so that it is curving in the opposite way. Look at the waves using the stroboscope.

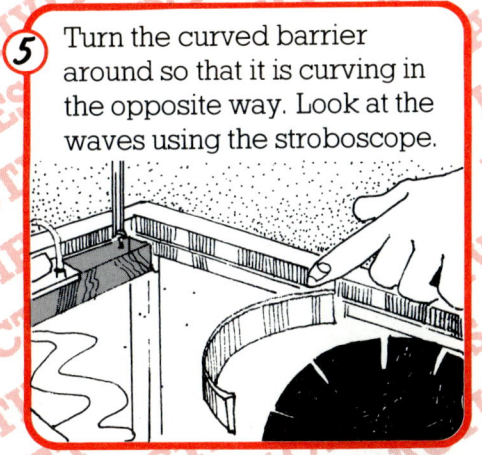

6. Repeat steps 4 and 5 using circular waves.

Answer these questions

1. How could you tell that the waves were reflecting?
2. Draw diagrams of the reflections you saw when the angle between the barrier and the waves was 45°.
   What can you say about the angle between the waves and the barrier compared with the angle between the reflections and the barrier?
   What would happen if you changed the angle of the barrier? Try it to see if your answer is right! Compare each angle with the angle between the reflections and the barrier. Do you notice anything? Can you think of a rule that describes the reflection of plane waves from flat barriers?
3. What happens to plane waves when they reflect from a concave barrier?
4. There is a 'special place' to start a circular wave near a concave barrier. Did you find it? If you did, draw what happened when the wave was reflected.

MAKING PATTERNS WITH WAVES

This Activity shows you three interesting effects that you can see using water waves. They are called diffraction, interference and refraction.

You will need
- ripple tank
- darkened room • lamp
- motor fixed to a beam
- rods for support
- dippers
- hand stroboscope
- power supply
- perspex sheet (about 0.5 cm thick)
- flat and curved barriers
- elastic bands

1 Diffraction

Set up the motor to make plane waves. Send them towards a gap between two small straight barriers. Use the stroboscope to see what happens to the waves after they pass through the gap. What you see is known as **diffraction**. Answer questions 1 and 2.

2 Interference

Put two dippers into the water, so that they just touch the surface. They should be about 2–3 cm apart. Switch on and look at the **interference pattern**. Answer questions 3 and 4.

3 Refraction

Gently submerge the perspex sheet in the water. Pass plane waves over it. Draw what happens to the waves. The changes you can see are called **refraction**. Answer questions 5 and 6.

Answer these questions

1 What happens to the diffraction pattern if you make the gap bigger or smaller?
2 Compare the wavelength of the waves used with the gap size that gives the *best* diffraction pattern. What do you notice?
3 What happens to the interference pattern when you change the separation of the dippers?
4 Make a note of the patterns you see. Can you explain how this interference pattern is caused?
5 What happens to the refraction when you change the *depth* of the water in the tank.
6 What does this experiment tell you about *light*? (Look at page 26 to remind yourself about the refraction of light.)

MEASURING THE SPEED OF SOUND

Sound waves move through the air much more *slowly* than light waves do. In fact, light moves so quickly that it will travel across several hundred metres almost instantly. In this Activity you can use this fact to find out how fast sound waves move.

You will need
- two metal dustbin lids
- long tape measure
- stopwatch

Copy this table

Distance /m	Time on stopwatch /s	Speed of sound m/s

1. Find a large open space, such as the games field. Measure the distance between two places at opposite ends. Try to get a distance of about 500 m if you can.

2. Stand at one of the places. Send your partner with the two bin lids to the other place. Ask him or her to bang the lids together when they get there.

3. *Start* the stopwatch when you *see* the lids come together. *Stop* the stopwatch when you *hear* the bang. Repeat this several times. Put the readings in the table.

4. Calculate the speed of sound for each reading

$$\text{speed} = \frac{\text{distance}}{\text{time taken}}$$

Write your results in the table.

Answer these questions

1. What was your average value for the speed of sound? (Find this by adding up all of the numbers in the last column of your table. Then divide that total by the number of readings.)
2. Do you think that your measurement of the time taken was accurate? If not, why not? How could the method be made more accurate?
3. Can you think of another example of light moving instantly and the sound coming shortly afterwards?
4. How could you measure the speed of sound through water?

LOOKING AT TRANSVERSE AND LONGITUDINAL WAVES

In this Activity you are going to look at the differences between longitudinal and transverse waves. You are also going to find out what is meant by **compression** and **rarefaction**.

You will need
- slinky spring
- small piece of paper
- sticky tape

1 Tell your partner to hold one end of the spring. Hold the other end so that it is slightly stretched out across the table. Your partner should keep his or her hand still during the Activity.

2 Quickly flick your wrist to the right of the rest position, then over to the left and back to the middle again. See what happens.

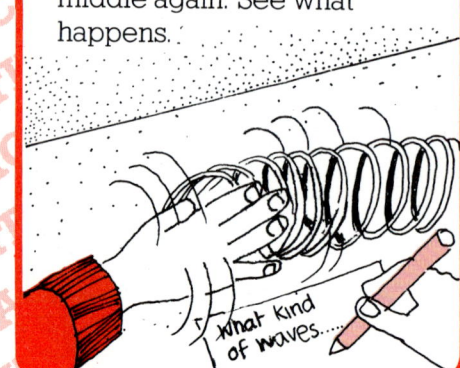

3 Stick a small piece of paper to one of the coils of the spring. Repeat step 2. See what happens to the piece of paper. Answer questions 1, 2 and 3.

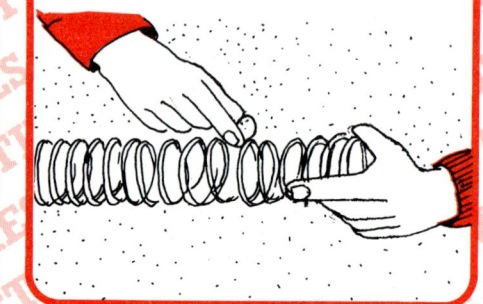

4 Hold the spring as before. Move your hand, sharply, 15 cm towards your partner, then move it backwards to about 15 cm away from the starting place. Then move it back to the starting place. See what happens.

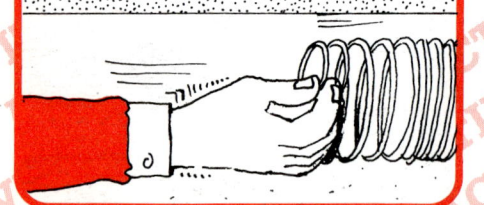

5 Repeat step 4 looking at the piece of paper.

6 Draw a diagram of this wave showing compressions (spring squashed together) and rarefactions (spring stretched out). Answer questions 4 and 5.

Answer these questions

1. What kind of wave moves down the spring?
2. What happened when it hit your partner's hand?
3. How did the piece of paper move?
4. Compare the position of the piece of paper before and after the wave had passed along the spring.
5. What actually went down the spring to your partner's hand?

MAKING SOUNDS WITH GUITAR STRINGS

Before you play a guitar you have to tune the different strings. You do this by stretching them in turn until they make the notes. If you want to make a different note on a tuned string, you change its length. This Activity will show you what effect these changes have.

You will need
- sonometer
- weights
- tuning fork
- block
- different wires

1 Put small weights on the end of the sonometer wire so that the wire 'twangs' when you pluck it.

2 Change the length of the wire by moving the bridge. Keep on plucking the wire. Listen to the note.

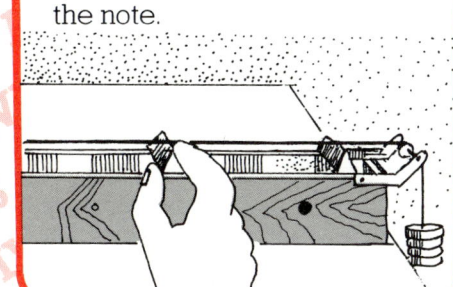

3 Change the wire. Repeat steps 1 and 2.

4 Keep the length of the wire the same. Change the weights. What effect do different weights have on the note?

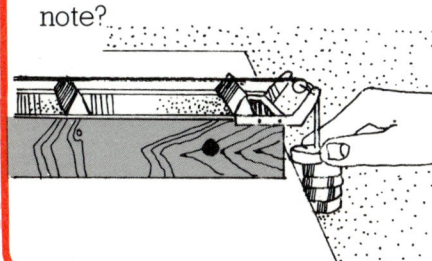

5 Strike the tuning fork on the block. Put the tuning fork base on the box.

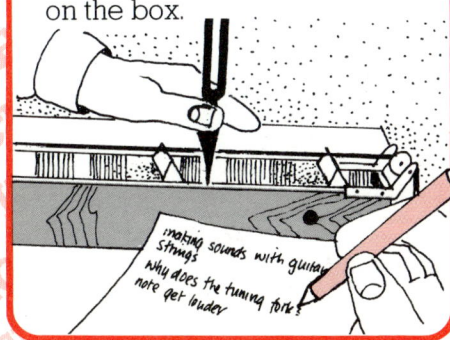

6 Try to match the note of the tuning fork with the note from the sonometer. Do this by altering the weights and the wire length.

Answer these questions

1. What happened to the note when you changed the length of the wire?
2. What happened to the note when you changed the wire?
3. What effect does changing the weight have on the note?
4. Why does the tuning fork's note get louder when the fork touches the sonometer box? (Clue: there are holes in the side of the sonometer box.)
5. Look at this guitar.
   (a) Which strings are used to give low notes?
   (b) How do guitarists use the frets? What do they use them for?
   (c) How do guitarists tune their guitars?
   (d) How are the notes from an acoustic guitar made louder?
   (e) How are the notes from an electric guitar made louder?

EXERCISES

1 Copy this out. Fill in the gaps.

There are two main types of wave, transverse and ………. When a ……… wave passes through water, the water moves up and down at 90° to the direction of movement of the wave. When a ……… wave passes through air, the air moves forwards and backwards in the same direction as the wave.

2 Answer these questions:
(a) Describe how you would set up a ripple tank to show a circular wave to a friend. Describe how you would 'freeze the action' using a stroboscope.
(b) How do you know that light can get through a vacuum but sound cannot?
(c) The musical note (A) has a frequency of 440 Hz. What frequencies would the two nearest octaves have?
(d) The quality of a musical note made by a trombone is different from the same note made by a guitar. Why?

3 (a) Describe the different causes of deafness. Which kinds can be helped with a hearing aid?
(b) How would your life be different if you were deaf? If you have a hearing difficulty describe the way it affects you.

4 Measure and write down the amplitude and wavelength of this wave motion.

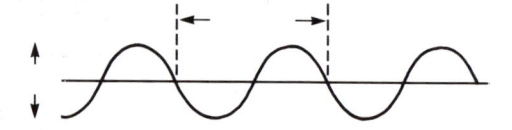

Now draw these three wave motions.

Amplitude/cm	Wavelength/cm
1	1
5	2
2	5

5 Find out the frequency of your local radio station and its wavelength of transmission.
 What do you think 'short wave radio' means?

6 These are two calculations involving the speed of sound.
(a) There is a flash of lightning and you begin to count. After 9 seconds you hear the rumble of thunder. If the speed of sound in air is 330 m/s, how far away was the lightning? Several seconds later there is a flash with only 6 seconds before you hear the sound of thunder. Is the storm coming nearer or going further away?
(b) HMS Pinger is out on routine patrol. The echo-sounder operator reports that the time interval between sending out and receiving a sound wave is 2 seconds. If the speed of sound in water is 1400 m/s, how deep is the sea?
 Suddenly, the time drops to 1 s. What could have caused this?

7 Compare the information on light in Chapter 2 with what you've just learned about refraction and reflection of water waves in the ripple tank. What does your comparison suggest?

8 Link up these words and statements

Ear-drum	Changes vibrations into nerve messages
Nerves	Contains three small bones
Middle ear	Carry messages to your brain
Oval window	Sense vibrations in the cochlea
Cells	Vibrates when sound waves hit it
Cochlea	Separates the middle ear from the inner ear

9 You hear sounds when the air is made to vibrate. List 10 sounds you have heard. Say what is making the air vibrate each time. Draw pictures to help your explanations.

10 Have you ever heard an echo? Describe it! Have you ever heard the Dopper Effect? When did it happen?

11 Find out and compare the range of frequencies you can make with your voice to those that bats make.

12 What is a compression? How is it different from a rarefaction?

4 MAGNETS AND ELECTRICITY

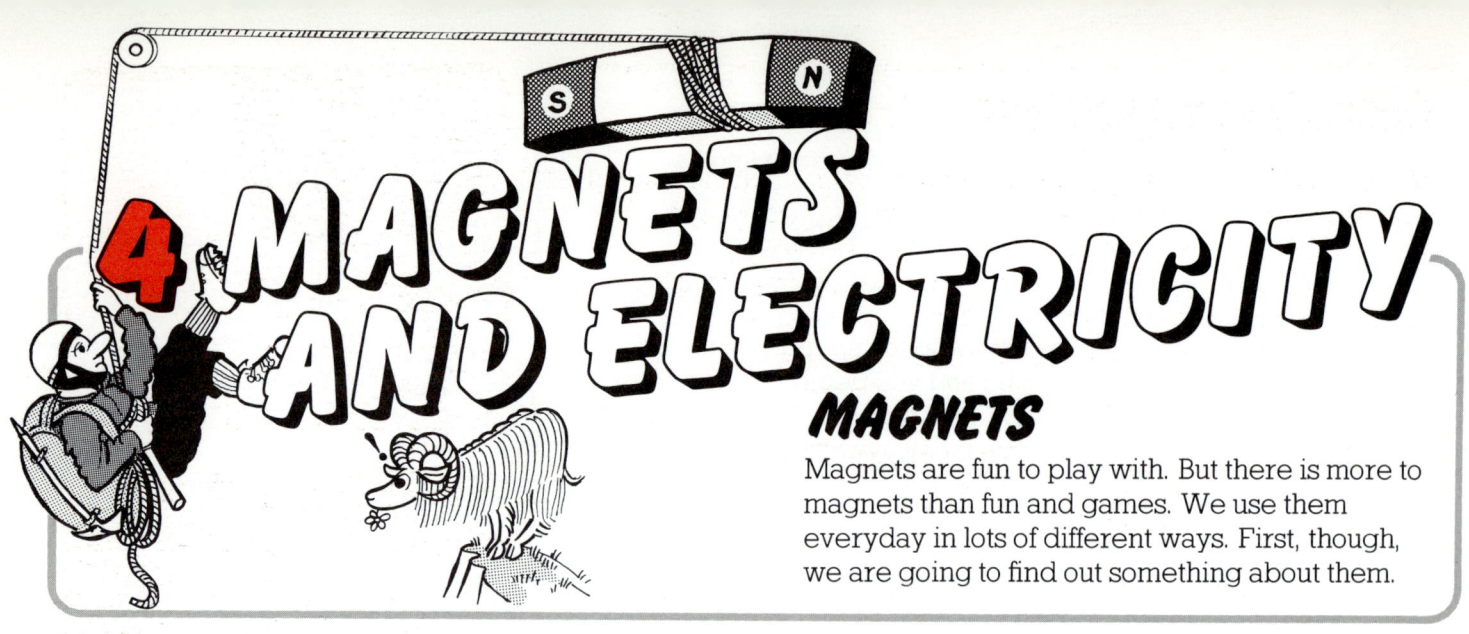

MAGNETS

Magnets are fun to play with. But there is more to magnets than fun and games. We use them everyday in lots of different ways. First, though, we are going to find out something about them.

ABOUT MAGNETS

If you hang up a magnet on a piece of cotton, it will always settle with one end pointing to the North, and the other pointing to the South.

The end pointing North is called the **North pole**. The end pointing South is called the **South pole**.

The strength of a magnet is concentrated equally at the poles. Look at these iron filings clinging to the ends of this magnet.

If you bring two South poles together, you will feel them trying to keep apart. We call this **repulsion**. You will feel the same thing with two North poles.

If you bring a North pole and a South pole together, you will feel them pull towards each other, we call this **attraction**.

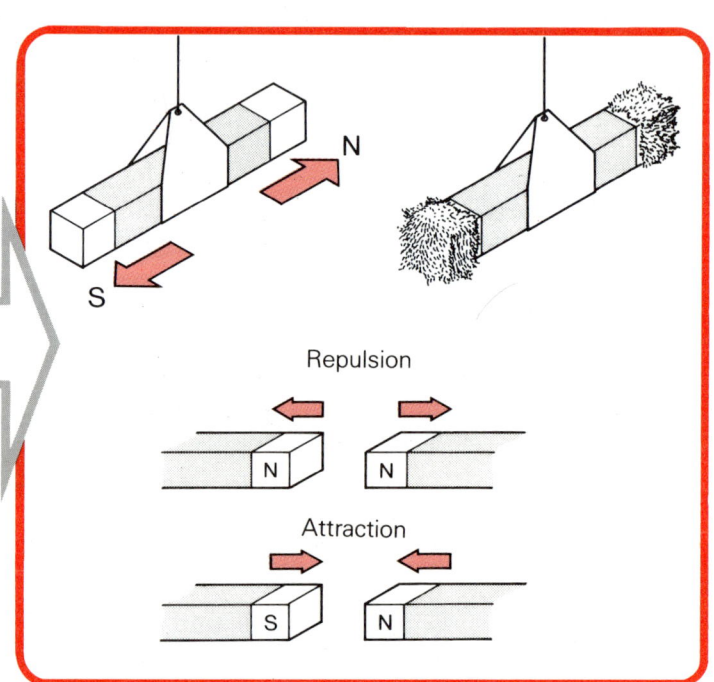

Repulsion

Attraction

Making a magnet

You can make magnets from pieces of iron, cobalt or nickel. These are called **ferromagnetic** materials. The simplest way to make a magnet is to use another one. To magnetise a nail you have to move the magnet along the nail, over the top, and back again about 50 times. In this picture the South pole would be at the point of the nail. The North pole would be at the head.

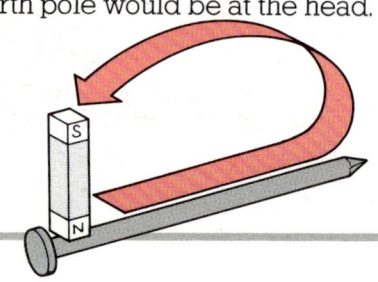

What happens in the nail?

Inside the nail, the atoms of iron form groups like little magnets. They are called **dipoles**. Each one has a North pole and a South pole. In an unmagnetised nail the dipoles point in all directions.

Rubbing a nail with a magnet moves the dipoles. It makes them all point in the same direction. The magnetism of the dipoles *inside* the bar cancels out. This leaves only the uncancelled Norths at one end and the uncancelled Souths at the other.

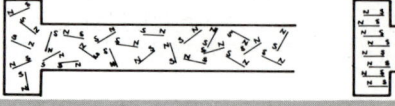

Inducing magnetism

When you use a magnet to magnetise a nail, we say that the magnet has **induced** magnetism in the nail. The same word is used to describe what's happened to the pins in this picture.

The pins become induced magnets. The magnetic poles of the pins are arranged so that they are attracted to one another.

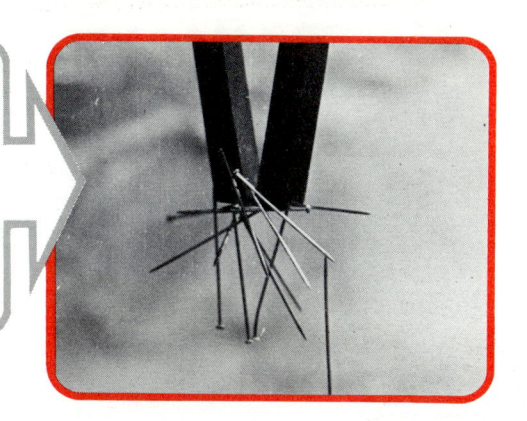

Magnetic fields

The space around a magnet which has magnetism in it is called the **magnetic field** of the magnet. You can find out how to map magnetic fields in the Activity on page 68.

A magnetic field surrounds a magnet in *all* directions. The iron filings are clinging to all sides of the horse-shoe magnet in this picture.

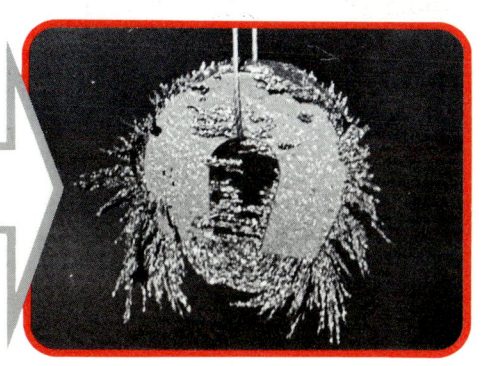

The Earth's magnetic field

The Earth has a magnetic field that reaches out tens of thousands of kilometres into space. It's as though the Earth had a giant bar magnet inside. (In fact, the magnetism is probably caused by the swirling action of the molten iron inside the Earth.)

If we could see the Earth and its magnetic field from the Moon it would look something like this.

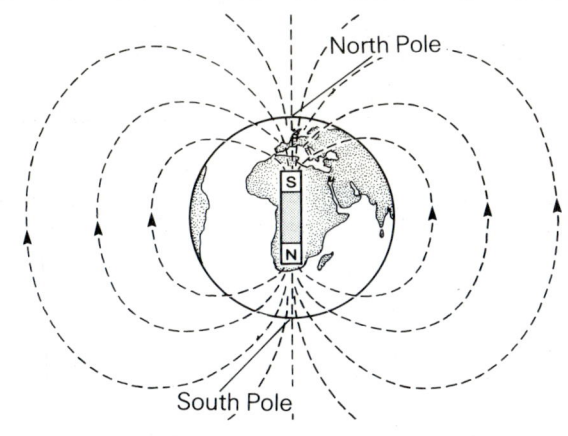

The lines only show the shape of the field. There are no lines really! The Earth's magnetic field passes through all of us.

You can see that the imaginary bar magnet inside the Earth has a South magnetic pole at the geographical North. This explains why the hanging bar magnet on page 56 points North. Opposites attract!

For centuries sailors have used compasses to help them to find their way. A compass is simply a small bar magnet. It is usually set up so that it can spin freely. If you leave it to settle, it will always end up pointing to the North.

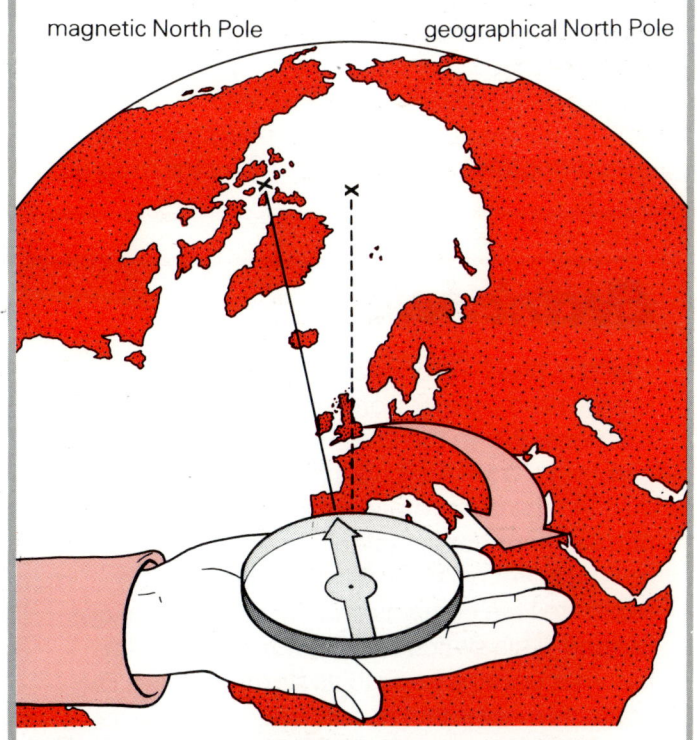

magnetic North Pole geographical North Pole

The magnetic pole is near the geographical North Pole.

57

ELECTROMAGNETS

Solenoids

Any wire with electricity passing through it has a magnetic field around it. It's difficult to believe because the field is so weak. But if you coil the wire up this concentrates the magnetic field in the wire. This sort of coil is called a **solenoid**.

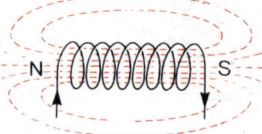

The field around the solenoid is exactly the same as the field around a bar magnet. It even has poles. You can see them marked in the drawing.

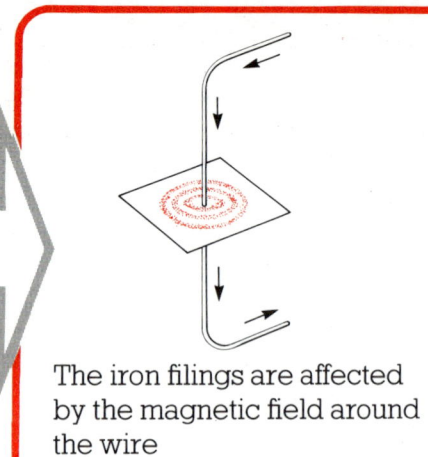

The iron filings are affected by the magnetic field around the wire

Switch on magnets!

You can use the field inside a solenoid to make an **electromagnet**. This is a magnet that you can switch on and off when you like. An electromagnet is a solenoid with an iron bar inside it. The electricity forces the dipoles in the iron bar to line up. This makes a very strong magnet.

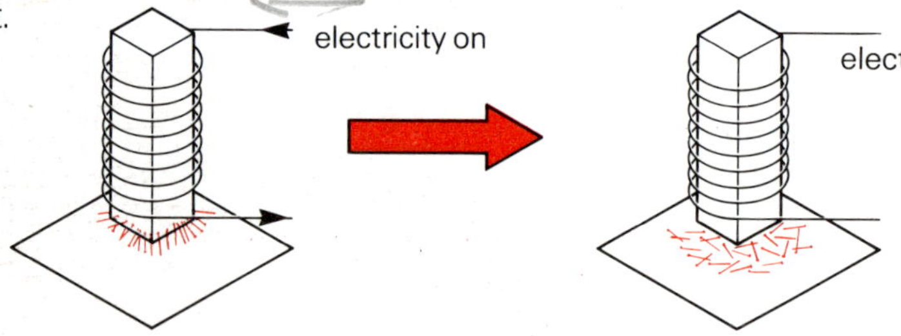

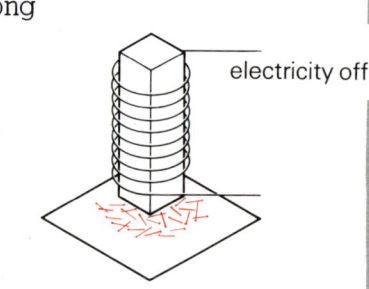

Switch on the current and the pins jump up to the electromagnet. Switch off, and the pins drop off.

Having a magnet that can be switched on and off is very useful!

Electric bells

Electric bells – perhaps like your school bell – are worked by electromagnets. This is what happens. When the timer switches to 'on', the electricity from the battery goes around the circuit. The electromagnet becomes magnetised and the hammer is pulled sideways to hit the gong. However, the sideways movement separates the contacts and the electricity stops flowing. This switches the magnet off and the hammer springs back to its starting place. The contacts meet again and the electricity flows, repeating the ringing action. This continues as long as the timer keeps the bell on.

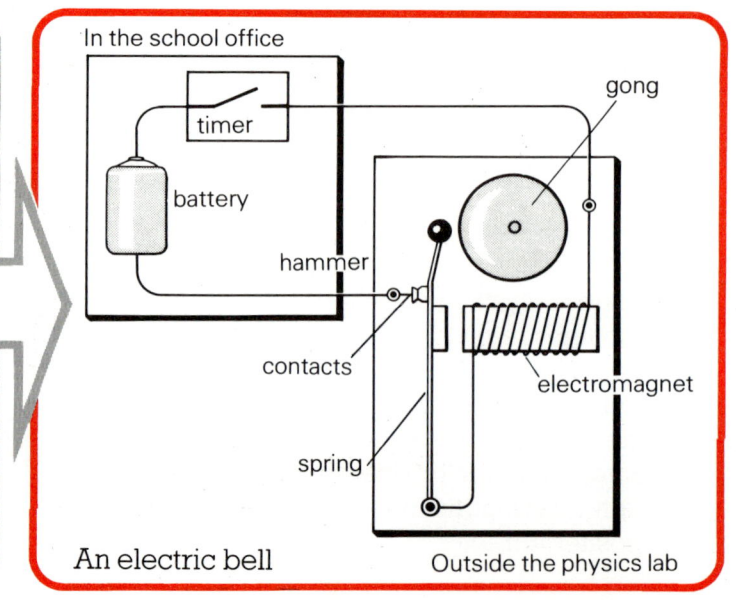

An electric bell

Outside the physics lab

Loudspeakers

Loudspeakers use a solenoid and a magnet to convert electrical waves into sound waves.

The solenoid is surrounded by the magnetic field from the magnet. One end of the solenoid is attached to the pointed end of the paper cone of the loudspeaker.

The 'music' comes to the solenoid in the form of a varying flow of electricity, called the **signal**. This electricity gives the solenoid a magnetic field. The field changes as the signal changes. The solenoid is either attracted or repelled by the permanent magnet depending on the signal. So the coil moves in or out. This moves the paper cone in or out, too. Movement of the cone squashes the air in front of it. This makes the sounds that you hear.

Electric motors use solenoids and magnets too. You can find out more about this in the Activity on page 70.

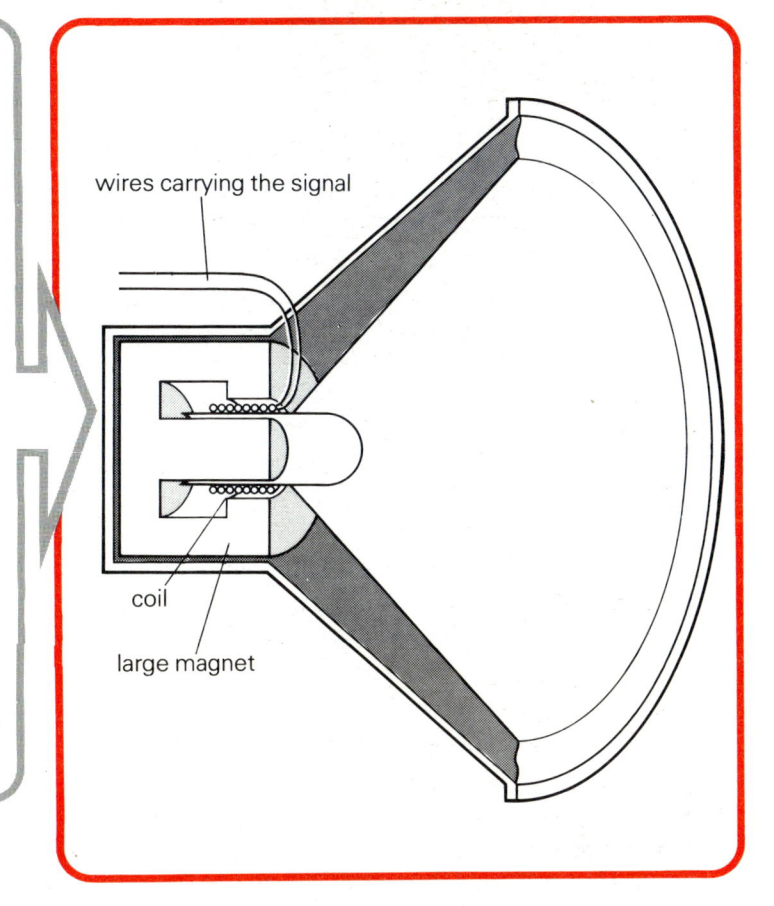

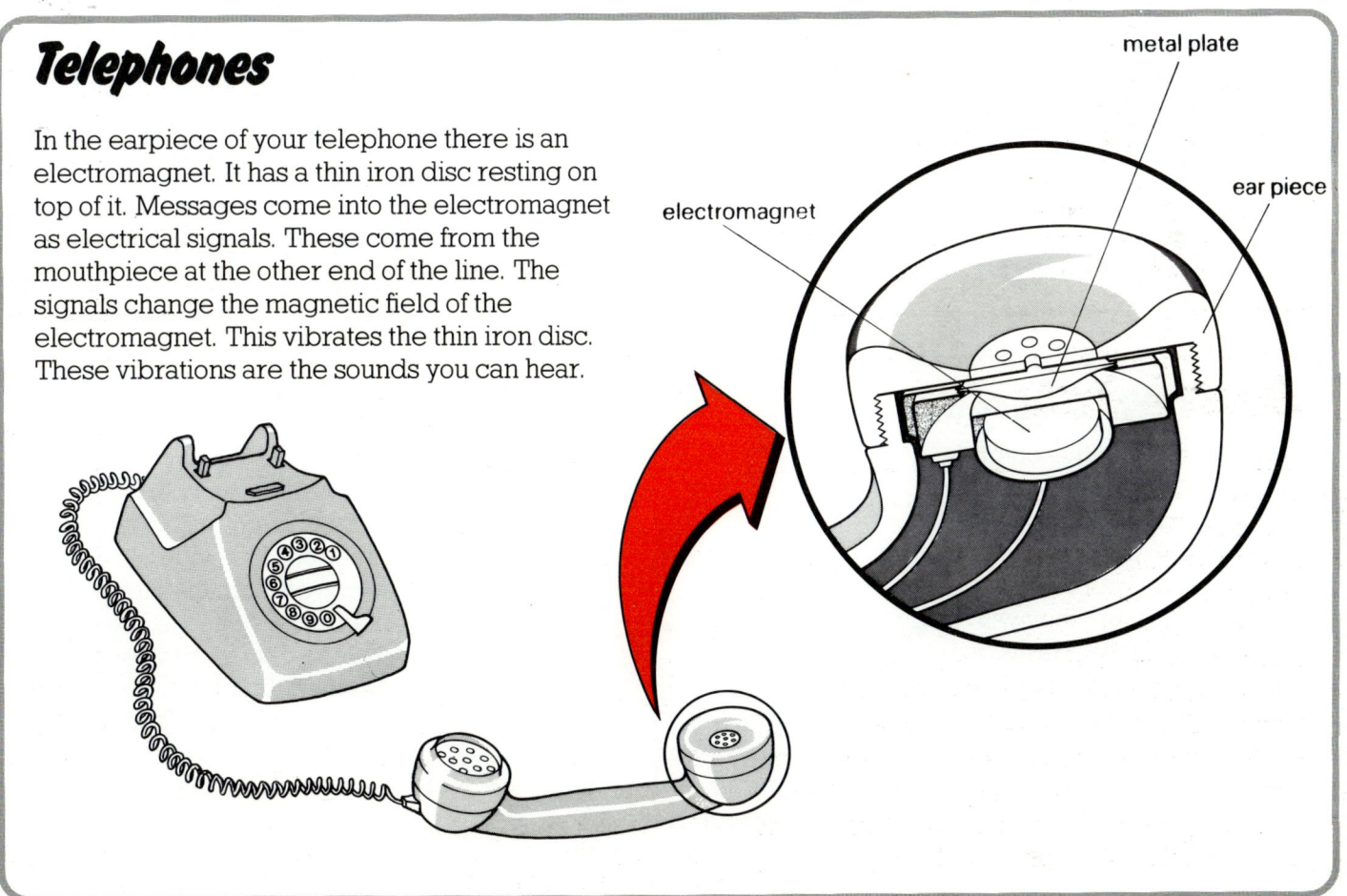

Labels: wires carrying the signal, coil, large magnet

Telephones

In the earpiece of your telephone there is an electromagnet. It has a thin iron disc resting on top of it. Messages come into the electromagnet as electrical signals. These come from the mouthpiece at the other end of the line. The signals change the magnetic field of the electromagnet. This vibrates the thin iron disc. These vibrations are the sounds you can hear.

Labels: metal plate, ear piece, electromagnet

STATIC ELECTRICITY

What is it?

Have you ever heard the crackle of tiny sparks when you take off a jumper? Have you felt an electric shock when you got out of a car, or when you touched metal after walking around on the carpets in a large department store? These are caused by **static electricity**.

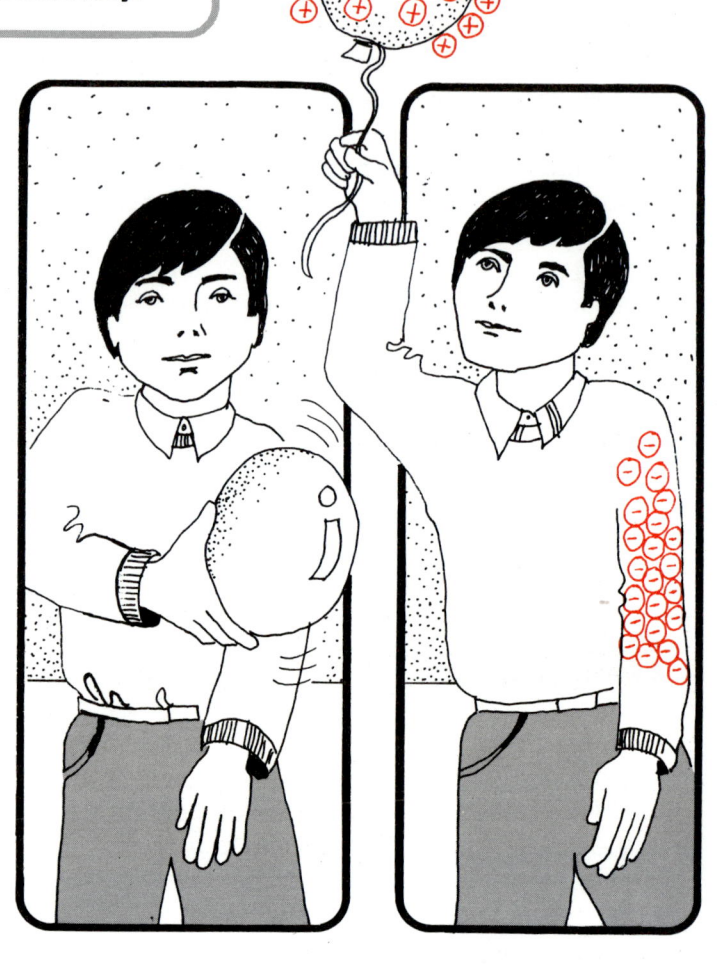

What causes it?

To understand static electricity you need to know something about atoms. Every atom has a **nucleus**. It has a positive charge (+). This nucleus is surrounded by many **electrons**. Each electron has a negative charge (−). Normally the number of positive charges is equal to the number of negative ones.

When you charge something, you either remove some of its electrons or give it extra ones. For example, you can charge a balloon by rubbing it on your woollen sleeve. When you do this you rub some of the electrons off the balloon by friction, and they stick on the wool. So the balloon has a positive charge and your sleeve has a negative charge.

You can charge plastic things easily. It's more difficult to charge metal things. This is because every metal is a good **conductor**. The electrons can easily escape through them to reach the Earth. Electrons cannot move through the plastic because it is an **insulator**.

Different charges

Positive and negative charges behave like magnetic poles. You can give two balloons the same charge by rubbing them separately on your sleeve. If you hang them up, they will repel one another. This is because they have the same charge.

If you charge two balloons by rubbing them together, they will stick to each other. This is because each balloon has the opposite charge. Electrons from each balloon have stuck to the other! Opposite charges attract.

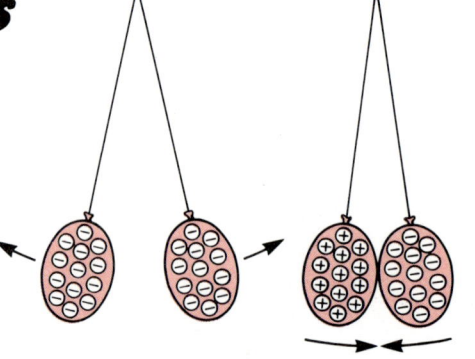

Lightning

The Earth is an enormous store of electrons. When electrons are added to or removed from this store to neutralise a charge, we say that the charge has been **earthed**. Lightning is simply a giant charge being earthed. This giant charge is caused by the water particles in a thunder cloud becoming charged as they move around.

Large static charges

The **van de Graaff generator** can produce large charges. When these are earthed they make large sparks that you can see easily.

The rubber belt is charged by friction at the base. The charges move up to the dome where they collect. The girl shares the charges with the dome. They cannot escape through the girl to Earth because she is standing on an insulator. Her hair is charged along with the rest of her body. Each hair has the same charge so they all repel each other. That's why it's such a hair-raising experience!

Electrostatic wind

Look at this picture. There is a drawing pin on the dome of the van de Graaff generator. When the generator is charged, the girl can feel a cool wind blowing up from the pin. Where does it come from?

Electrons gather on the dome and are especially crowded at the pin's point. This charge is very strong. It can repel any charged air molecules nearby. The wind the girl can feel is actually molecules rushing away from the charge.

Lightning conductors

A lightning conductor is a strip of copper attached to a tall building. Look at tall buildings around your school and you will probably see one. It has a spike at the roof end. The other end is attached to a metal plate buried in the ground.

A thundercloud has an enormous charge on it. If the charge is negative, the lightning conductor works like this. As the cloud passes over the conductor, electrons are repelled through the plate into the Earth. A positive charge is left on the spike. Air molecules with positive charge rush away from the spike. They partly neutralise the charge in the cloud. This makes a lightning strike less likely.

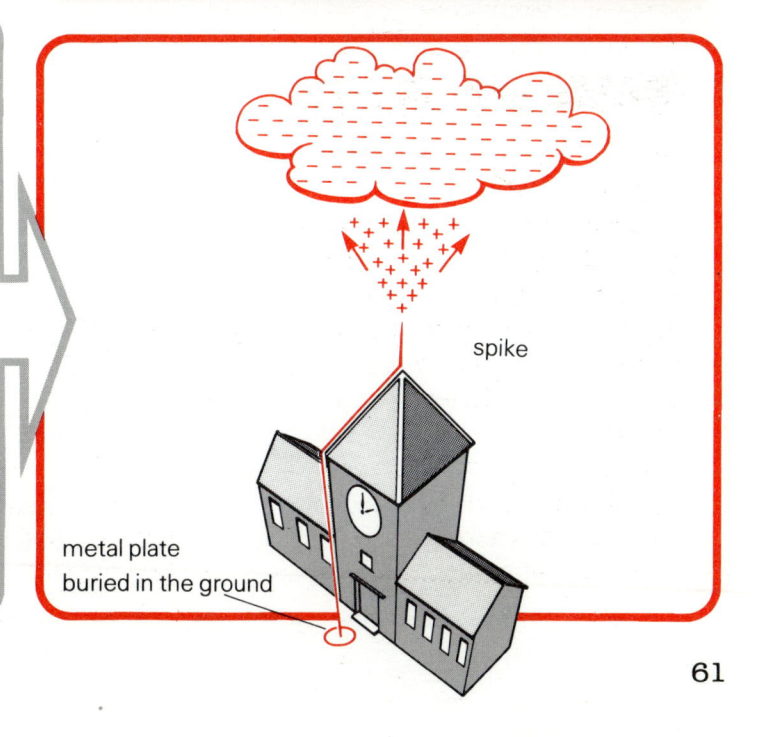

MOVING CHARGES – ELECTRICAL CURRENT

a.c. and d.c.

When electrons move along a wire, we say an **electric current** is flowing. There are two kinds of electric current: **direct current (d.c.)** and **alternating current (a.c.)**.

You can 'see' which is which by using an oscilloscope. It converts the current into a trace on a screen.

When the battery is connected in this simple circuit, the bulb lights up as the current passes round. The oscilloscope trace shows that the electron flow is steady and all in the same direction. This is direct current (d.c.).

If the connections on the battery are reversed the trace looks like this.

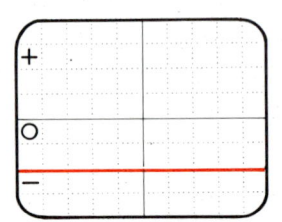

This is because the electrons are flowing in the opposite direction.

Now the battery is replaced by a bicycle dynamo.

When the dynamo is turned the bulb lights as the current passes round the circuit. The oscilloscope traces out how the electrons flow. They move backwards and forwards as the handle is turned. This is alternating current (a.c.).

You will see that the bulb lights up just the same. You can't see just by looking at the light bulb whether the current is direct or alternating. The effect produced in the bulb is the same!

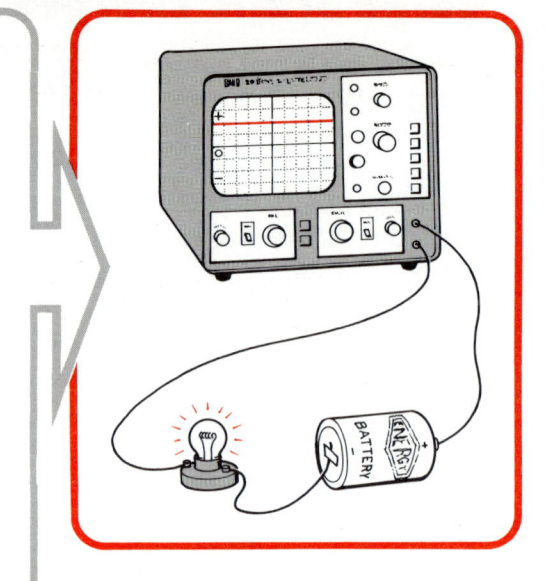

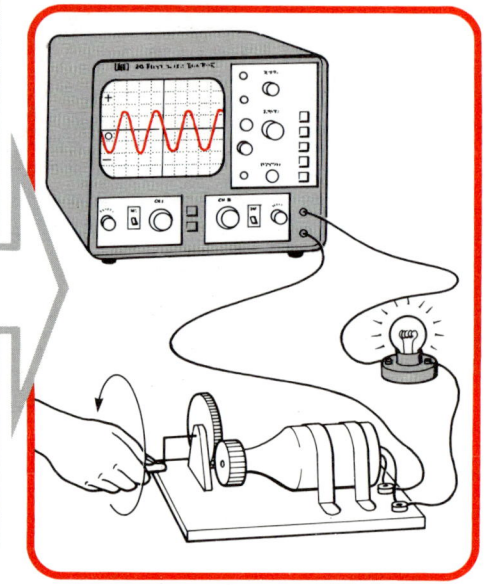

Mains electricity

Your mains electricity supply at home is alternating current. The current changes direction one hundred times a second. If you could see it slowed down by an oscilloscope it would look like this.

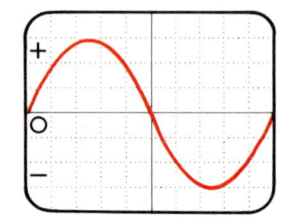

You would see 50 of these cycles every second. We say that mains electricity has a frequency of 50 hertz (look at page 40 to find more out about frequency).

Ammeters

You can measure the size of an electric current with an **ammeter**. One kind of ammeter works like this.

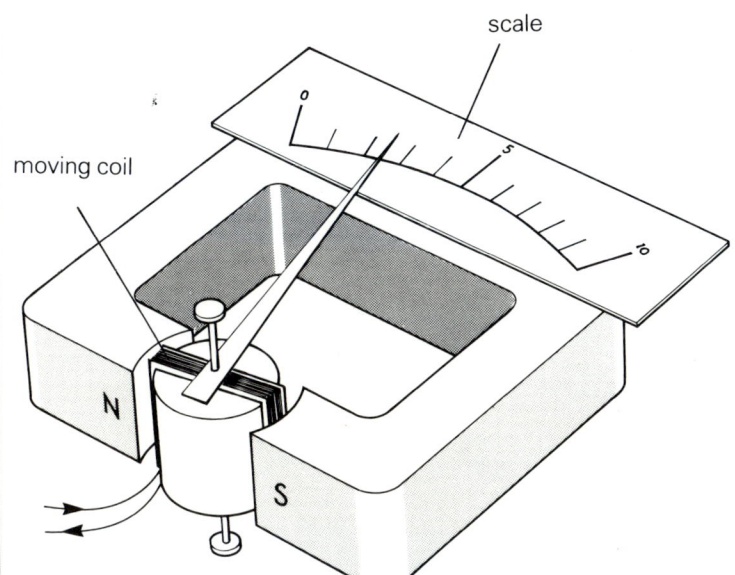

scale

moving coil

Inside the meter is a coil surrounded by a specially shaped magnet. When electricity goes through the coil, it makes a magnetic field. Bigger currents make bigger magnetic fields. The magnetic field of the magnet moves the coil until the magnetic force is balanced by the force of a spring fastened to the coil. The pointer shows the size of the current. The bigger the current, the further the needle moves.

This meter gives a steady reading with direct current only. An alternating current would make the pointer flicker backwards and forwards around the zero mark. You have to use a special meter to measure the size of an alternating current.

Current is measured in **amperes** (A). This is often shortened to **amps**.

Potential difference

The amount of 'push' needed to get current flowing from one place in a circuit to the next is called the **potential difference**. It is measured in **volts** (V). You need another kind of meter to measure a potential difference between different places in a circuit. It is called a **voltmeter**. (It works in much the same way as the ammeter, but with slight modifications.) Look at this circuit. The voltmeter is measuring the potential difference needed to get the current through the second bulb. The battery provides 9 volts of 'push' altogether. The ammeter is measuring the current passing through the bulbs. You can use an ammeter and a voltmeter to find out more about circuits in the Activity on page 74.

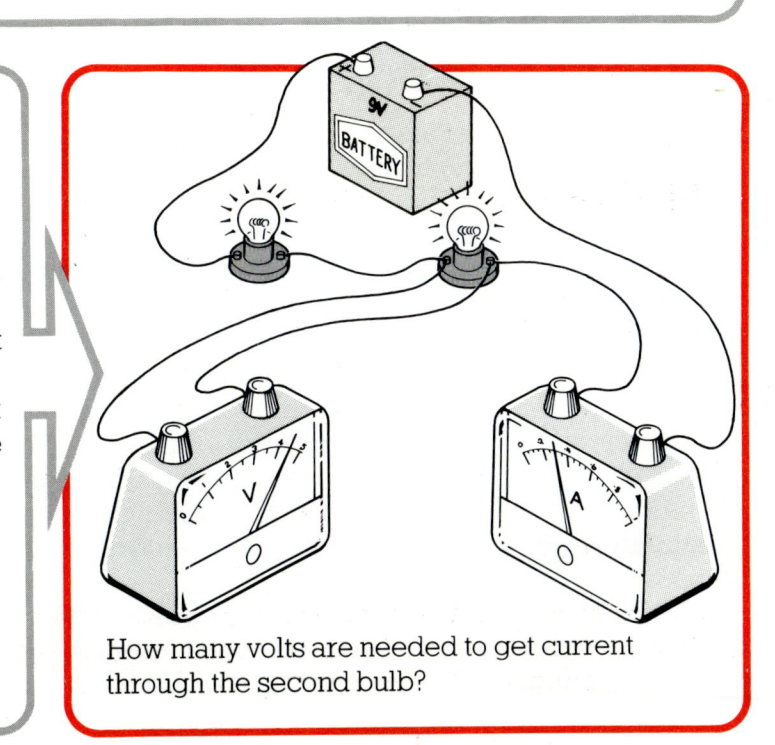

How many volts are needed to get current through the second bulb?

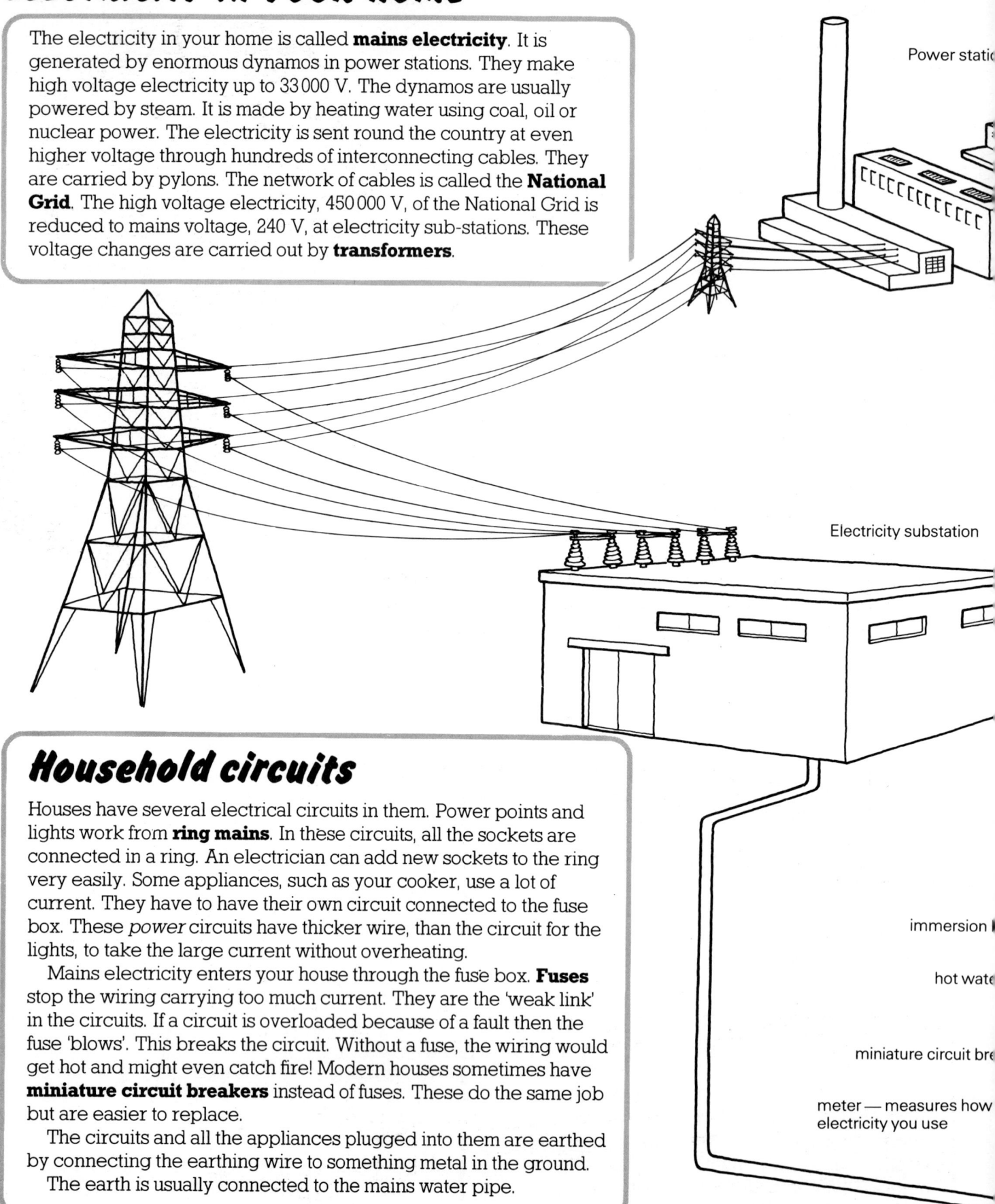

ELECTRICITY IN YOUR HOME

The electricity in your home is called **mains electricity**. It is generated by enormous dynamos in power stations. They make high voltage electricity up to 33 000 V. The dynamos are usually powered by steam. It is made by heating water using coal, oil or nuclear power. The electricity is sent round the country at even higher voltage through hundreds of interconnecting cables. They are carried by pylons. The network of cables is called the **National Grid**. The high voltage electricity, 450 000 V, of the National Grid is reduced to mains voltage, 240 V, at electricity sub-stations. These voltage changes are carried out by **transformers**.

Power station

Electricity substation

immersion

hot wate

miniature circuit bre

meter — measures how electricity you use

Household circuits

Houses have several electrical circuits in them. Power points and lights work from **ring mains**. In these circuits, all the sockets are connected in a ring. An electrician can add new sockets to the ring very easily. Some appliances, such as your cooker, use a lot of current. They have to have their own circuit connected to the fuse box. These *power* circuits have thicker wire, than the circuit for the lights, to take the large current without overheating.

Mains electricity enters your house through the fuse box. **Fuses** stop the wiring carrying too much current. They are the 'weak link' in the circuits. If a circuit is overloaded because of a fault then the fuse 'blows'. This breaks the circuit. Without a fuse, the wiring would get hot and might even catch fire! Modern houses sometimes have **miniature circuit breakers** instead of fuses. These do the same job but are easier to replace.

The circuits and all the appliances plugged into them are earthed by connecting the earthing wire to something metal in the ground. The earth is usually connected to the mains water pipe.

earth (yellow and green) — fuse

neutral (blue)

live (brown)

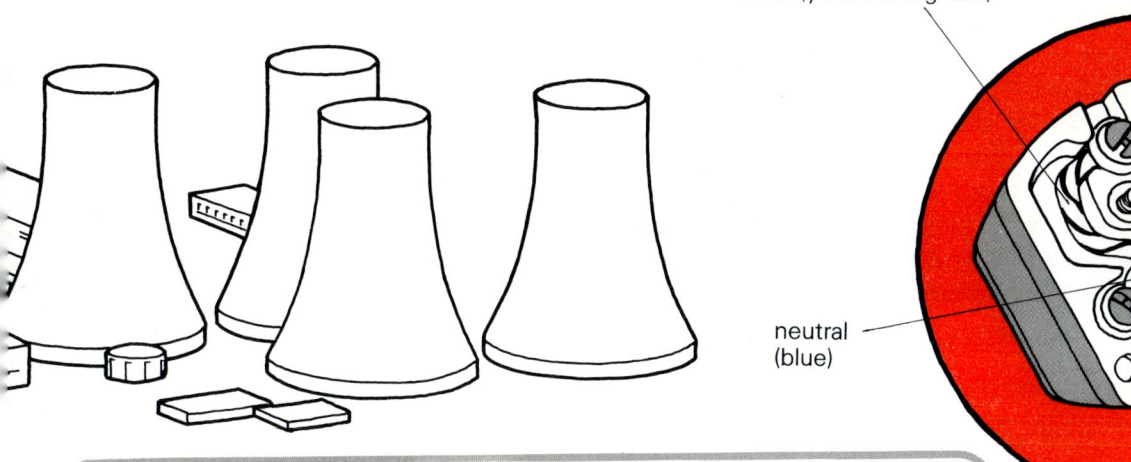

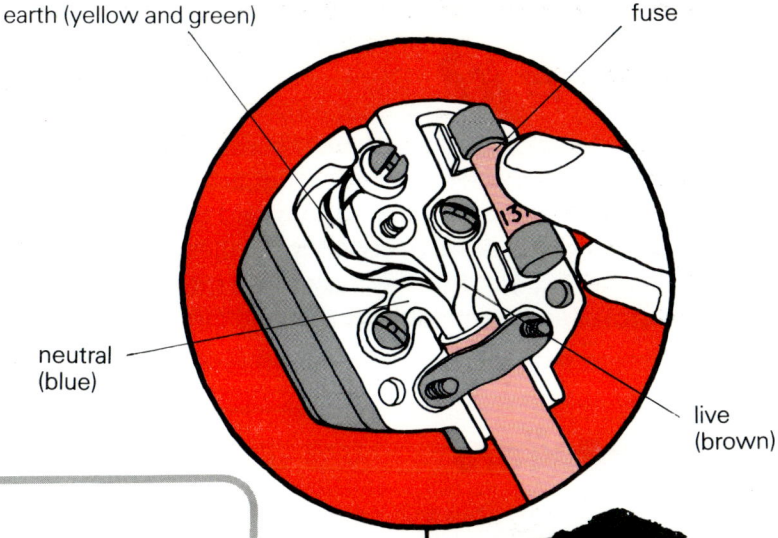

Plugs and power points

Each appliance you use has to be connected to the power points on the ring main by a plug. This too has a fuse. It protects the thing you are using, not the house.

The mains electricity comes into the plug through the live (brown) wire. It comes back out of the plug through the neutral (blue) wire. The earth wire connects the metal case of the appliance to earth. This is an important safety measure. Think what would happen if the live wire became disconnected inside the appliance and touched the case. If the case was earthed, the electricity would be taken to earth straight away. If the case wasn't earthed, you would get an electric shock when *you* touched the case. You must always make sure that electrical gadgets with metal parts which you can touch are properly earthed.

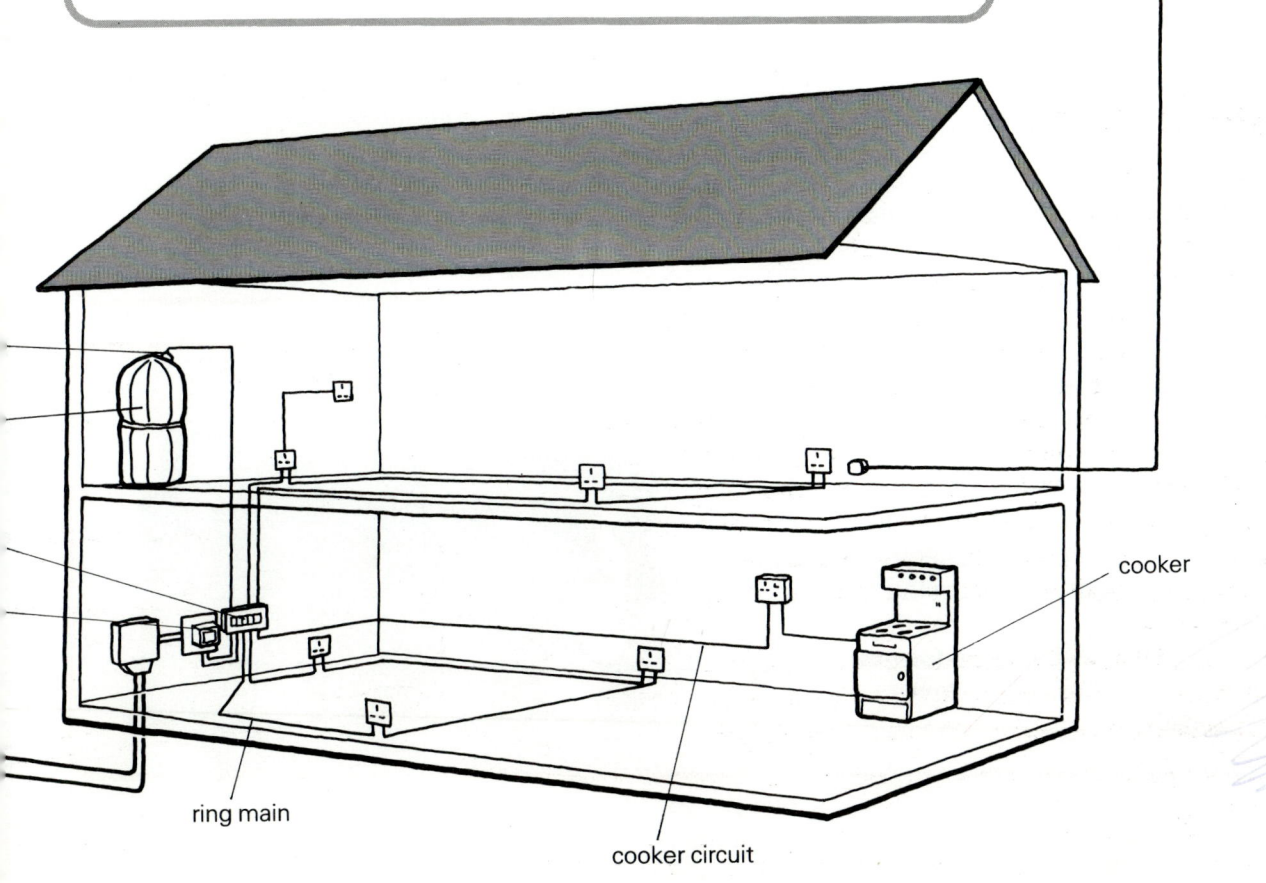

Warning: never use a plug with the back cover removed

cooker

ring main

cooker circuit

ELECTRONICS

What is it?

Many electrical gadgets in your home, such as the TV, radio and stereo, contain the same components. The most important of these are **resistors**, **capacitors**, **diodes** and **transistors**. They are connected in electrical **circuits**. All these components control the movement of electrons as they travel through those circuits. This control of electron movements in the circuits is called **electronics**.

Capacitors

Capacitors are able to hold electrical charge. This picture shows you the inside of one kind of capacitor. There are two thin metal plates separated by a layer of insulator. This sandwich is rolled up to save space.

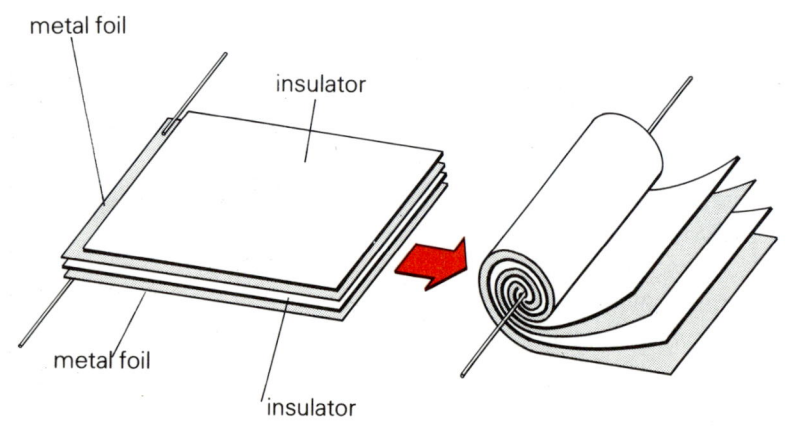

metal foil
insulator
metal foil
insulator

If charge builds up on one of the plates, then the opposite charge is induced on the other plate. The insulator increases the ability of the capacitor to hold the charge. It stops electrons jumping from one plate to the other. Any change in current on one side of the capacitor will produce an equal change on the opposite side. So, a capacitor will *pass on* **changes** *in current but not a continuous current*.

Resistors

Resistors reduce the current flowing in the circuit by resisting the flow of electrons. Most resistors have a fixed value of resistance measured in **ohms**. Ω is the sign for ohms. You can find out more about ohms in the Activity on page 74.

This is a variable resistor. It is sometimes called a **potentiometer**. It can give a range of resistance simply by turning the metal sliding contact.

Changing the voltage

This photograph shows the inside of a radio. You can see the different electrical components.

Like many of the electrical gadgets in your home, this radio uses *low voltage d.c.* electricity. This comes either from batteries or from the *high voltage a.c.* mains supply. The high voltage is converted to a low voltage by the **transformer**. The alternating current is changed to direct current using a **rectifier**.

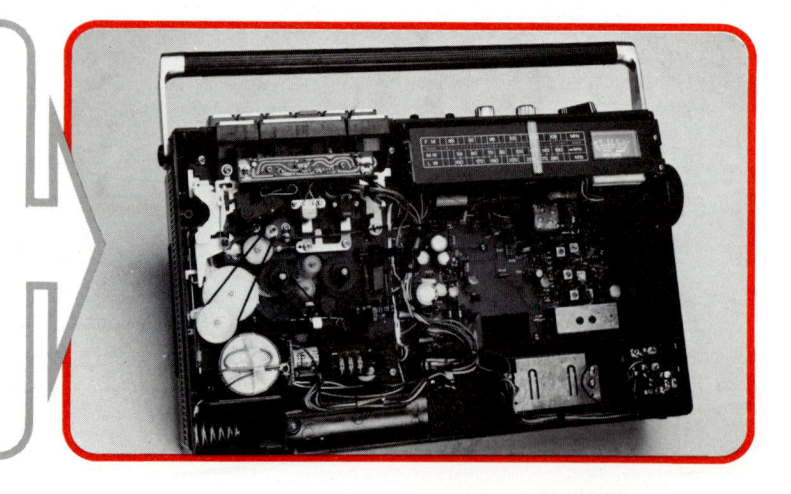

66

Diodes and transistors

Diodes and transistors are made from **semiconductors**. These are elements with unusual conducting properties. Silicon is a semiconductor. Its ability to conduct electricity can be changed by adding a small amount of a different element to it. This is called **doping**.

One type of diode is made from a small chip of silicon doped with phosphorus and boron. Doping with these two elements means that current can flow only in one direction through the chip.

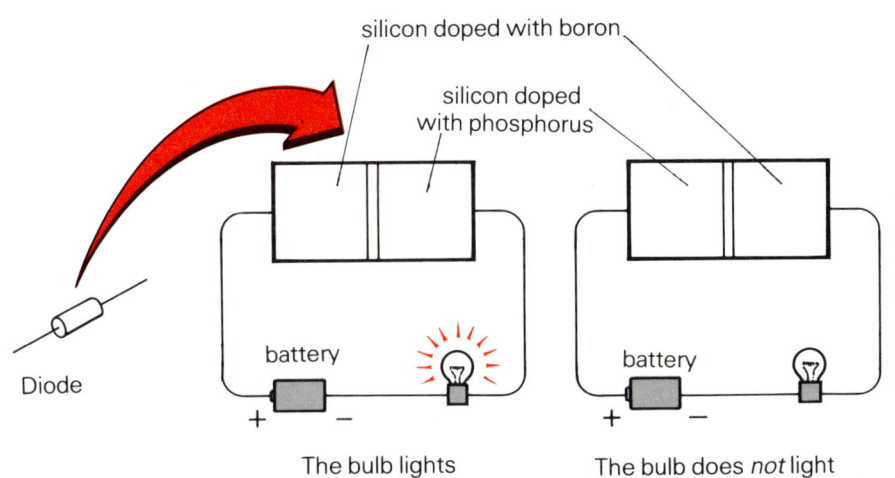

The bulb lights — The bulb does *not* light

Transistors, like diodes, are made from semiconductors. A transistor has *three* sections though. These sections link two separate circuits so that the current flowing through one of the circuits controls the current flowing in the other.
1 If there is no current in the first circuit then there can be no current in the second circuit. A transistor used in this way acts like a **switch**.
2 A small change in the current in the first circuit produces a large change in the current in the second circuit. A transistor used in this way is acting as an **amplifier**.

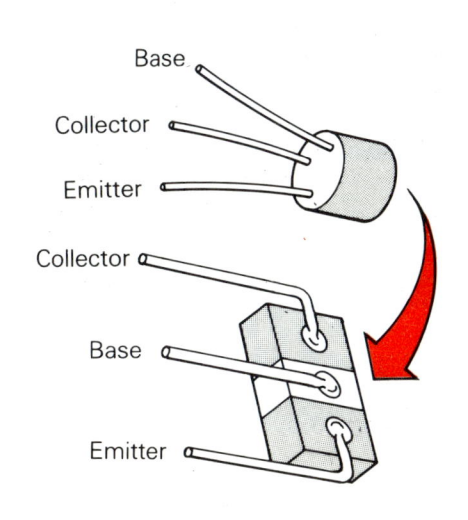

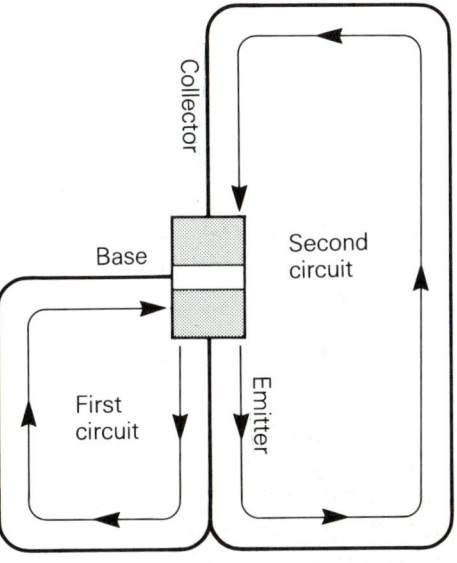

Integrated circuits

Integrated circuits can contain many transistors, diodes, capacitors and resistors set into a single **silicon chip**. The chip is designed and then made smaller photographically. This tiny design is *printed* onto the silicon. Different areas are doped or coated or etched according to the design. This forms the **components**.

A **microprocessor** has many integrated circuits set into its silicon chip. One microprocessor can have tens of thousands of circuit components! Integrated circuits can be programmed to do specific tasks. For example, they can be calculators, make TV video games or control washing machines. They form the basis of computers, robots and word processors.

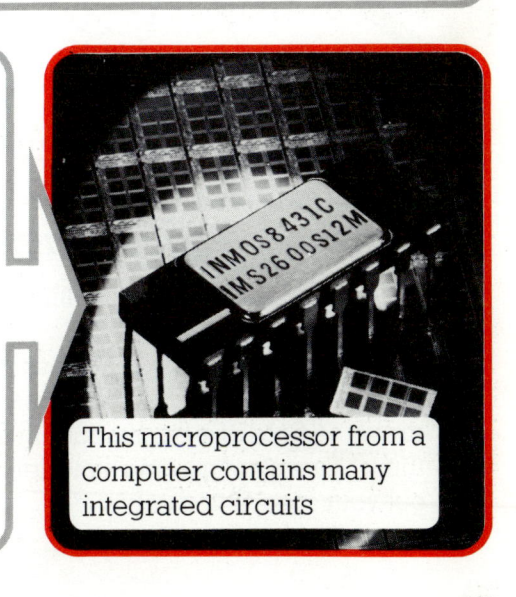

This microprocessor from a computer contains many integrated circuits

FINDING OUT ABOUT MAGNETIC FIELDS

If you move a nail towards a magnet you can feel the magnetism before you touch the magnet. This is because of the **magnetic field**.

In this Activity you are going to find out about the shapes of magnetic fields that surround magnets. Iron filings will map out the magnetic fields for you. They will become induced magnets themselves. So, they will be attracted to the poles of the magnet you are studying.

You will need
- different magnets
- iron filings ● paper
- wax ● tripod
- large saucepan
- Bunsen burner
- heat proof mat

1 Carefully melt the wax in the large saucepan. Dip the paper in. Let it dry. This will coat the paper with a thin layer of wax.

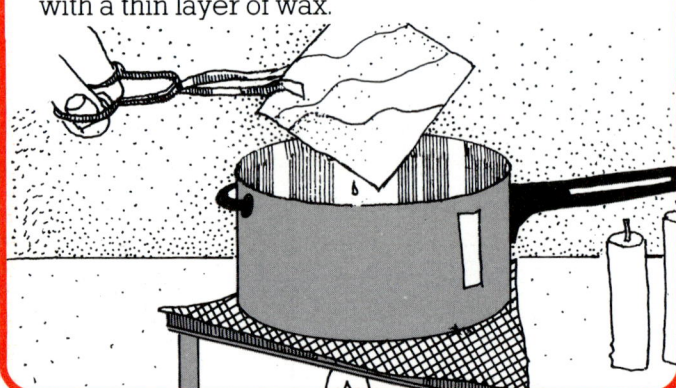

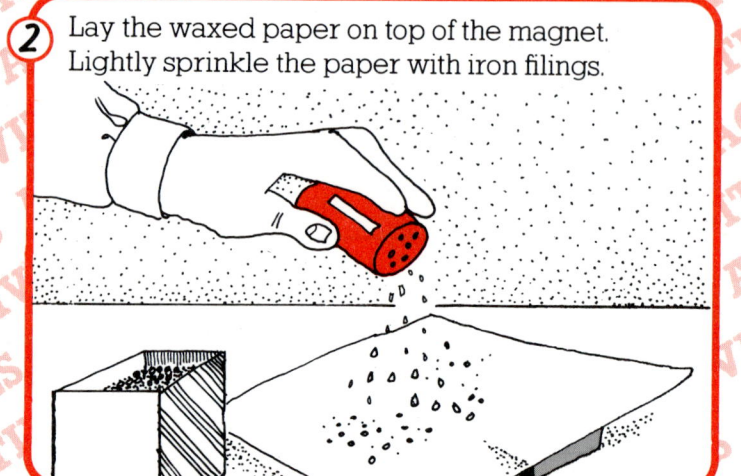

2 Lay the waxed paper on top of the magnet. Lightly sprinkle the paper with iron filings.

3 When you are happy with the pattern, adjust the Bunsen burner to get a *small* blue flame. *Gently* play the flame over the magnetic field. This will melt the wax.

4 Let the wax set. You will now have a permanent record of the magnetic field. Repeat steps 1–4 with other magnets and with pairs of magnets close to each other.

Answer these questions

1 Where did most of the iron filings gather? Explain why this happened.
2 The magnetic fields you have studied are all 3-dimensional. This means that they occupy space *all around* the magnets. Try to draw the field of a bar magnet in a '3-dimensional' way. Can you think of a way of making a model of the whole magnetic field? Try out your idea, if you can.

ELECTROSTATICS

These Activities will help you find out about electrostatics. Electrostatic charge can make strange things happen! The effects are all caused by the **induction of charge**.

You will need
- polythene strip
- perspex strip
- paper
- balloons
- wool
- soap solution

(1) Rub the polythene strip on the wool. The strip will be negatively charged. Hold it close to a thin stream of water coming out of a tap. What happens to the water? Predict what will happen if you repeat this with a *positively* charged perspex strip. Try it.

Why does it happen?

The negative charge on the polythene repels the electrons in the water molecules. The side of the water nearer the polythene becomes positively charged. This positive charge is attracted to the negative charge on the polythene, so the water bends towards the polythene. This is called **electrostatic induction**.

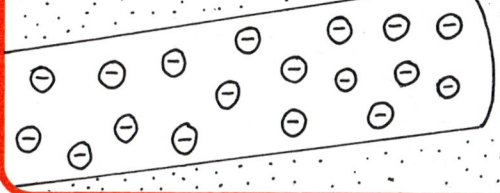

(2) Charge a balloon by rubbing it on your jumper. Make sure you rub in one direction only. If you rub backwards and forwards you will replace the electrons that you removed. Try to stick the balloon to the wall.

Why does it happen?

When you rub the balloon you give it a positive charge. If you hold it near the wall it induces a negative charge on the wall. So the balloon and the wall are attracted to each other.

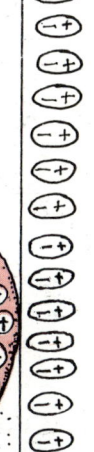

(3) Rub the polythene strip on the wool. Hold the polythene near a soap bubble. Look carefully to see what happens.

Use the explanations of what happened in steps 1 and 2 to work out what happened this time.

69

MAKING A MOTOR

An electric motor is a gadget that converts electricity into movement. The electricity runs through a coil of wire which sits in between two magnets. This makes a force that turns the coil.

You will need

- reel of insulated copper wire
- steel U-shape
- 2 Magnadur magnets
- knitting needle
- wooden block with a tube through the middle
- 2 split pins
- sticky tape
- wooden base
- 4 studs
- power pack

1 Experiment with the Magnadur magnets. Find out where the poles are.

2 Fix the Magnadur magnets onto the steel U-shape. Make sure their opposite poles face one another across the gap.

3 Wrap sticky tape around the ends of the thin tube that runs through the wooden block.

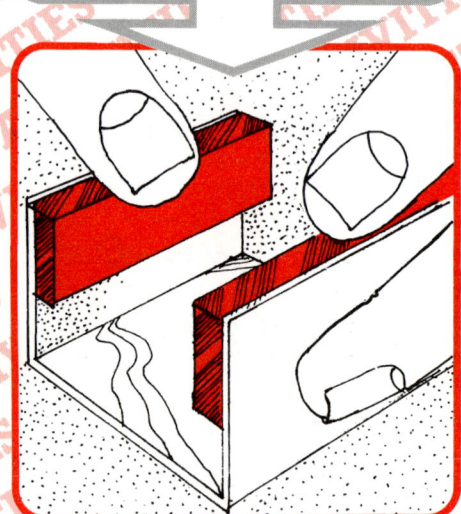

4 Wrap about thirty turns of wire around the wooden block. Start and finish with about a centimetre of bare wire.

5 Hold the bare ends of the coil wire in place on the taped tube with small elastic bands.

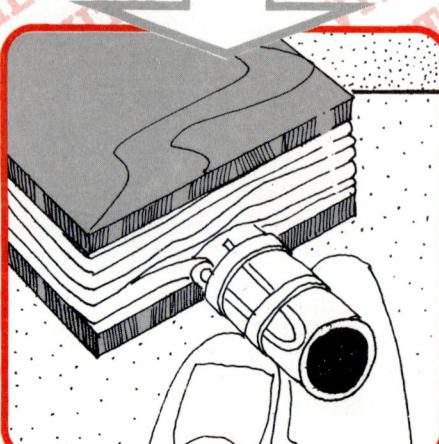

6 Stick the split pins in the wooden base. Position the coil between the magnets. Put the knitting needle through the tube and through the split pins.

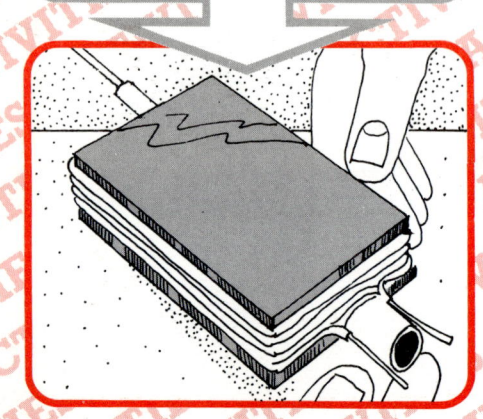

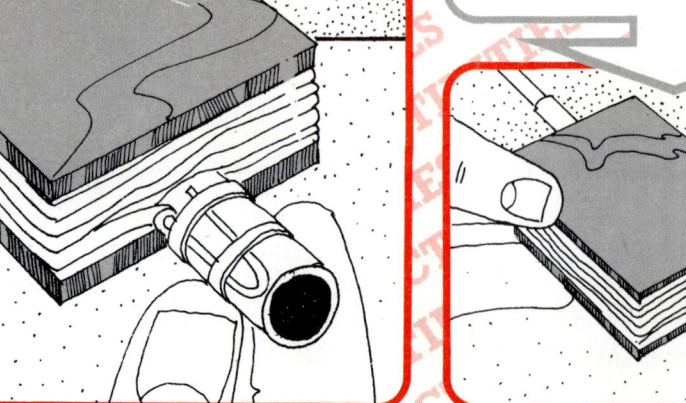

7 Cut two 20 cm lengths of insulated wire. Strip off two centimetres of insulation from one end of each.

8 Arrange the bare ends to brush against the ends of the coil wire. Hold these brushes in place using the studs. Press the studs into holes on the base.

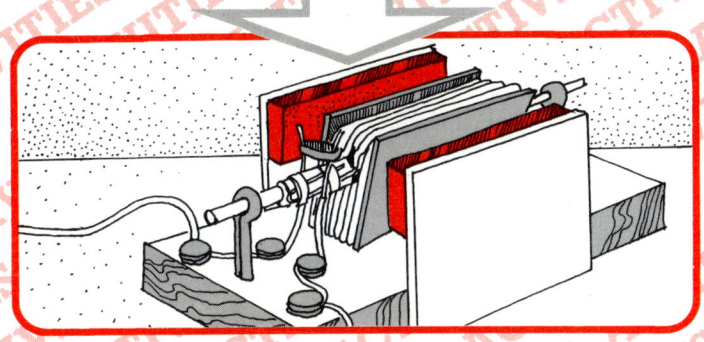

9 Slip another small elastic band over the top of the brushes. This will improve their contact with the coil wire. Take care to see that the brushes don't touch each other.

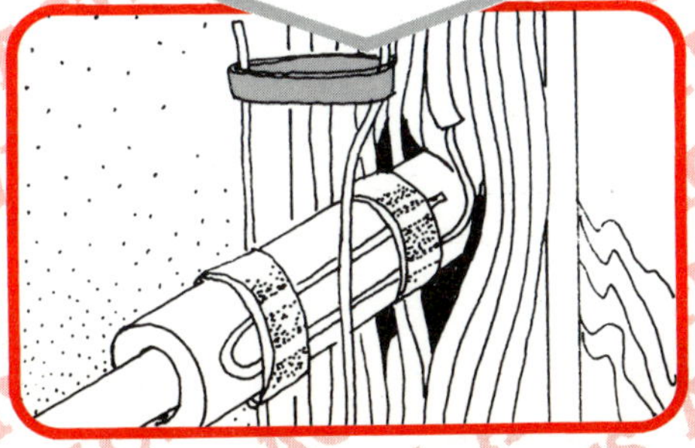

10 Connect up the other ends of the brushes to the power pack. Set it to 2 volts d.c. Switch on. Give the block a start and it should spin!

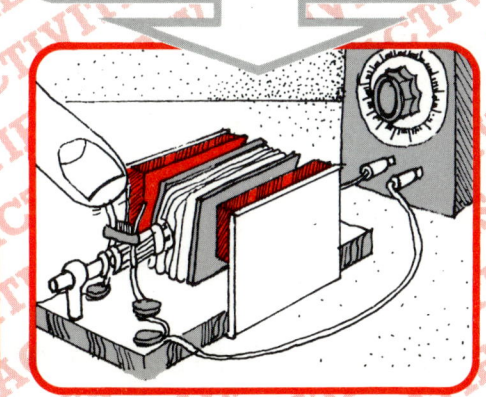

Answer these questions

1. What do you think will happen if you use a *slightly* higher voltage? Try it.
2. What happens if you reverse the contacts?
3. What would happen if you used a.c. instead of d.c.? Try it.
4. Try to work out how the motor keeps spinning.

How does it work?

Electricity flows in at the left-hand side of the coil and out of the right-hand side. This makes the upper side of the coil the South pole and the lower side the North pole. So, the coil turns in an anti-clockwise direction attracted by the Magnadur magnet poles.

In the second position there is no electricity entering the coil. But the coil's momentum keeps it spinning.

As the coil has gone round half a turn, the contacts with the brushes have swopped over. Electrical contact is made again. The upper side of the coil is still a South pole and the lower side a North pole.

The anti-clockwise movement continues, first because of repulsion but then, as the coil moves round a little further, by attraction. The whole cycle then starts again!

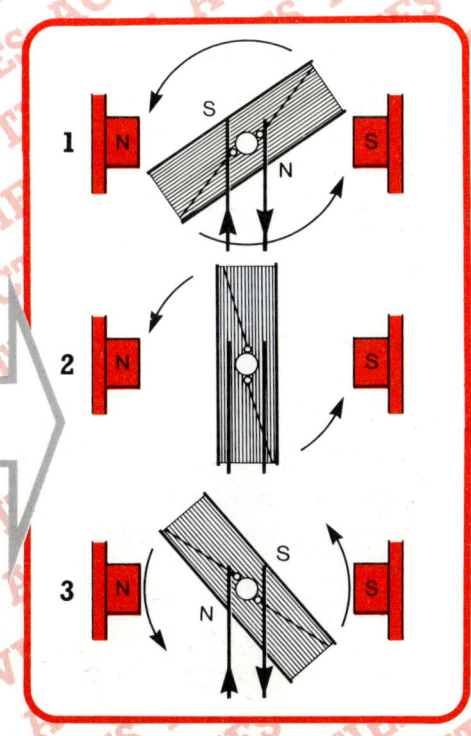

ELECTRICITY FOR FREE?

This Activity shows you how to make electricity in a simple way using a **chemical reaction**. You are going to make a battery, but from very unusual materials! You will use a **milliammeter** to measure the current your battery makes. This is a very sensitive meter and *you must ask your teacher to show you how to use it.*

You will need

- selection of fruit and vegetables including a lemon
- selection of strips of different metals such as a brass drawing pin and a paper clip etc.
- torch bulb in a holder
- milliammeter
- 2 leads with crocodile clips

Copy this table

Name of fruit or vegetable	Metal	Did the bulb light?	How much current was produced?

1 Cut the lemon in half. Stick the brass drawing pin in one side and the paper clip in the other. Connect the other ends to the bulb. Does it light? Replace the bulb with the milliammeter. Measure how much current is flowing in the circuit. Fill in the table.

2 Repeat step 1 with other fruits and vegetables. Try different metals as well. Put your results in the table.

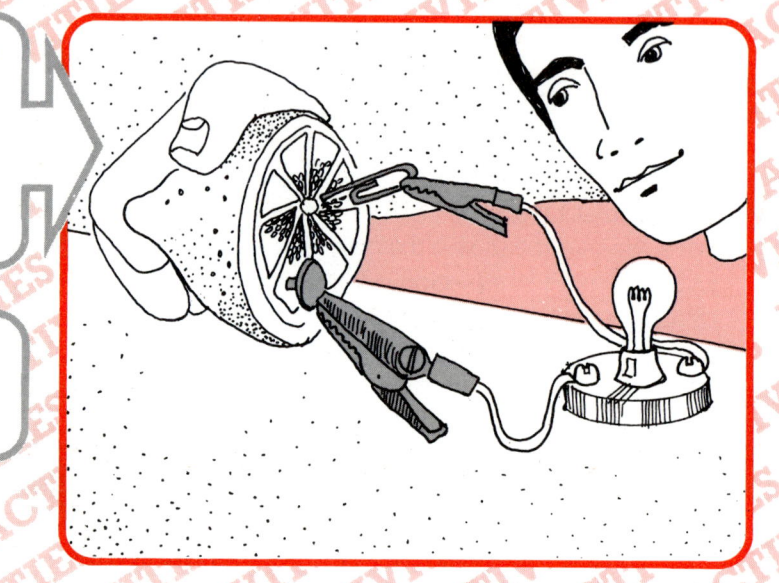

There is a chemical reaction in the lemon. As a result, electrons are sent around the circuit. The acid of the lemon is called the **electrolyte**. It can conduct the flow of electricity.

Answer these questions

1 Which combination produced the highest current?
2 Which combinations failed to produce any current at all?
3 What would you expect to happen if the metals were the same? Try it.

How a battery works

In a torch battery the electrodes are the **carbon rod** down the centre and the **zinc case** around the outside. The **ammonium chloride** jelly is the electrolyte. When the battery is connected in a circuit, the zinc slowly dissolves into the electrolyte. This releases electrons which run around the circuit to reach the carbon rod. This electron flow is the flow of electricity. The manganese dioxide and carbon mixture is called the depolariser. It prevents the build up of gas bubbles around the carbon rod by absorbing the gas. A battery without a depolariser would only work for a minute or so!

WHAT IS ELECTROMAGNETIC INDUCTION?

This Activity shows you how to make your own electricity without batteries! If you move a wire through a magnetic field an electric current flows down the wire. We call this **electromagnetic induction**. It is the basic idea behind the generation of electricity. The mains electricity you use in your home is made in this way. Bicycle dynamos work this way too.

You will need

- 30 cm piece of straight wire
- coil with a 100 turns of wire
- powerful horseshoe magnet
- bar magnet
- milliammeter

1 Connect the ends of the long wire to the milliammeter. Move the wire between the poles of the horseshoe magnet. Keep an eye on the meter. Move the wire in different ways to find out which movement gives the biggest deflection of the meter.

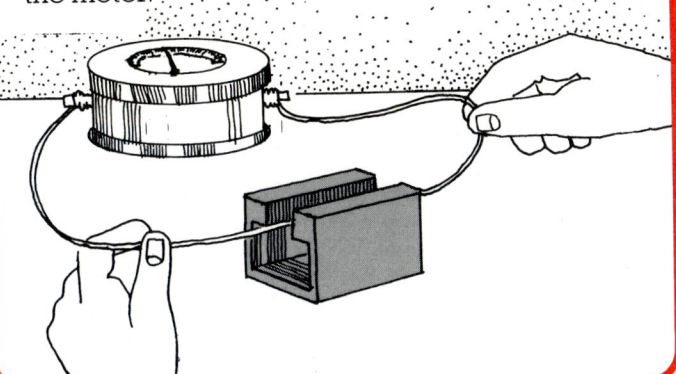

2 Disconnect the wire. Connect up the coil to the milliammeter. Move the bar magnet in and out of the coil. Look at the meter as you do this. Push the magnet in quickly, and then slowly. Pull the magnet out quickly, then try slowly. Swop the poles of the magnet round and try this again. Keep the magnet still and move the coil.

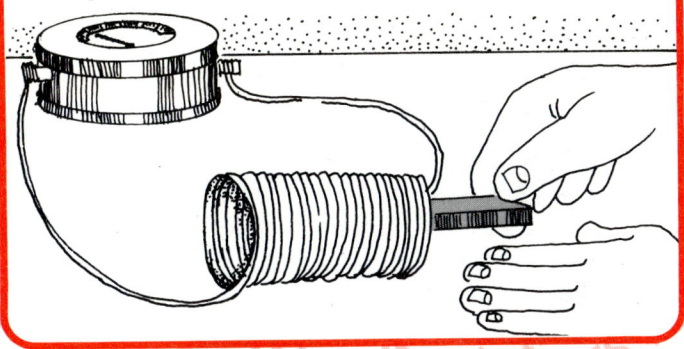

How does a bicycle dynamo work?

A bicycle dynamo works by electromagnetic induction. The knob turns as the bicycle wheel is pedalled round. Inside the dynamo, the knob is attached to a magnet. This magnet is next to a coil which is connected to the lamp. Electricity is induced in the coil as the magnet turns. The electricity flows through the circuit to light the lamp.

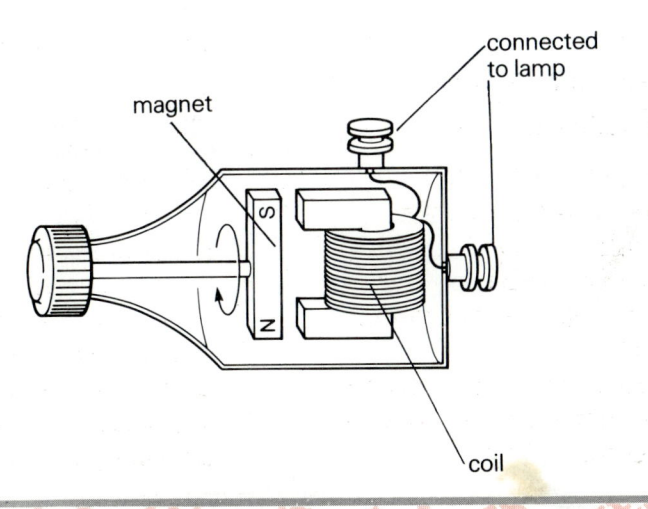

Answer these questions

1 What did the milliammeter measure?
2 In step 1, which movement made the most current flow down the wire.
3 In step 2 which movement made the most electricity? Which movement made the least?
4 What was the effect of swopping the poles of the magnet around in step 2?

RESISTORS

In a simple circuit, the battery provides the push which forces the current round. In some parts of the circuit, such as the bulbs or resistors, quite a bit of push is needed to get the current through. The amount of push needed is called the **potential difference**. It is measured in **volts** using a **voltmeter**. In this Activity you will be able to find out how the potential difference across the resistor affects the current flowing through it.

You will need

- ammeter (0–1 A d.c.)
- voltmeter (0–15 V d.c.)
- power pack
- resistors, 5.6 Ω, 10 Ω, 22 Ω or similar
- connecting wires
- bulb in holder

Copy this table

Potential difference across the resistor/volts	Current flowing through the resistor/amps

1. Connect up the power pack, ammeter and resistors like this.
   Use the d.c. output of the power pack. *Do not switch on yet.* The ammeter will measure the current flowing through the resistor.

2. Connect the voltmeter to the two ends of the resistor. The voltmeter will measure the potential difference across it.

3. Turn the power pack to zero. Switch on. Gradually increase the current up to 1 amp. Take pairs of readings of potential difference across the resistor and readings of current passing through the resistor. Write your results in the table.

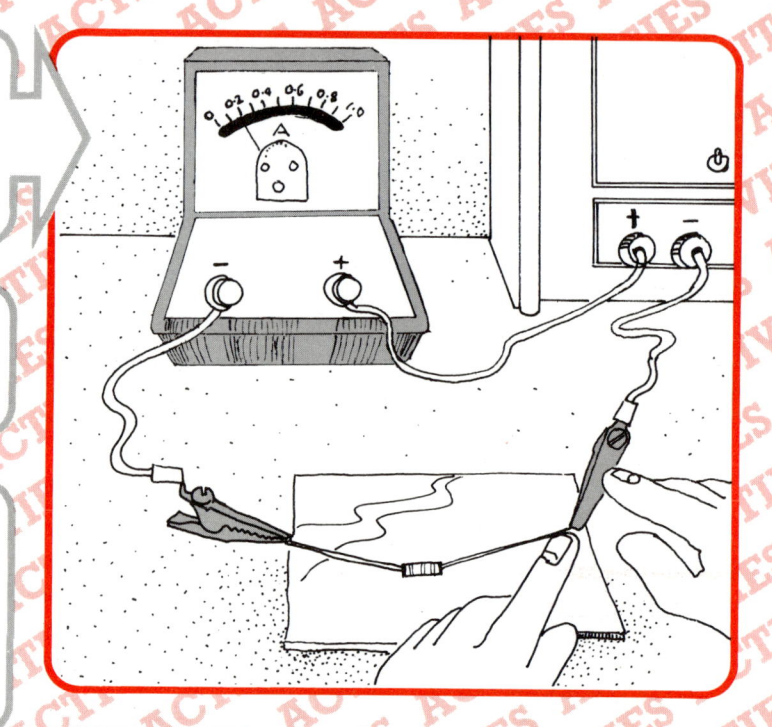

4. Plot a graph with potential difference (in volts) on the y axis and current (in amps) in the x axis.

Answer these questions

1. What does the shape of the graph tell you about the way in which the potential difference across the resistor affects the current flowing through it?
2. Repeat the experiment, if you can, with a different resistor. Plot your results as a different line on the same graph. Predict what shape the graph will be before you try the experiment.
3. Try replacing the resistor with a bulb. Plot the results on a different graph. Try to explain what you found.

FUSES

This Activity shows you how the fuses keep the wiring and the electrical gadgets in your house safe.

You will need

- 1 A and 5 A fuses
- power supply
- ammeter
- 5.6 Ω resistor
- heat-proof mat
- safety screen
- safety glasses
- connecting leads
- crocodile clips

1 Set up this circuit. Use the d.c. supply. The resistor represents an electrical gadget – a record player for instance. The 1 A fuse is there to protect it.

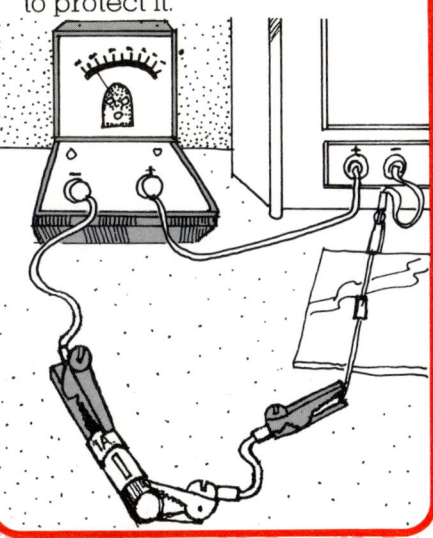

2 Turn the power pack to zero. Switch on. Gradually increase the power supply. Try to watch the fuse and the ammeter at the same time. Make a note of the ammeter reading when the fuse burns out. Switch off.

3 Replace the burnt-out fuse with a 5 A fuse. Place the resistor on the heat proof mat. *Put on safety glasses and set up the experiment behind the safety screen.* Turn the power pack to zero. Switch on. Gradually increase the power supply. Note what happens.

Answer these questions

1. How did the fuse value compare with the current at which it burnt out in the first experiment?
2. How did you know that the resistor was not protected in step 3.
3. What was the highest ammeter reading you reached in step 3.
4. What would happen in a house if the electrical gadgets were not protected with a fuse?

Fuses in plugs

Modern plugs have fuses in them. They act as safety devices to prevent too much current flowing down the live wire. Too much current might flow if the live wire came into contact with a metal part inside the electrical appliance. Without a fuse, the wiring or the appliance itself could be damaged. This picture shows the inside of a plug.

You can see where the fuse is fitted. It is important to use the correct one. An appliance rated at more than 700 watts – like a kettle or heater – needs a 13 amp fuse. A table lamp or a record player rated at less than 700 watts needs only a 3 amp fuse.

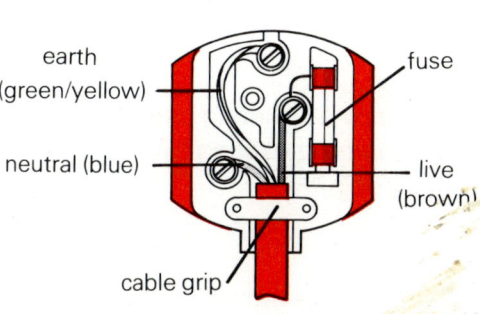

earth (green/yellow)
neutral (blue)
fuse
live (brown)
cable grip

75

EXERCISES

1 Copy this out. Fill in the missing words.

Magnets have two ...., a .... and a .... Like poles .... and .... poles repel. So, if you brought together the South pole end of one bar magnet and the North pole end of another, you would feel a force .... them together. On the other hand if you try to force two .... poles together, you will feel them .... apart.
Choose from: pushing, pulling, magnetism, unlike, attract, poles, North, South.

2 Match up the parts of these sentences:

The strength of a magnet	is like that of a bar magnet.
There are three connecting wires on	a negative charge.
The magnetic field around a solenoid	a transistor.
Electrons carry	alternating current.
a.c. stands for	is concentrated at the poles.

3 Give an example of what is meant by magnetic induction. What is electrostatic induction? Draw diagrams to make both of your answers clear.

4 Describe the difference between alternating current and direct current. What sort of device makes a.c.? What makes d.c.?

5 You are given two identical small bars of iron and told that one is a magnet. How can you find out which one is a magnet using only a piece of thread?
How could you have told them apart if you hadn't the piece of thread, or anything else to help?

6 Draw a lightning conductor. How does it help to prevent lightning striking? Where does the energy in a lightning strike go to?

7 What is a transistor made from? How can it be used to work (a) like a switch and (b) as an amplifier.

8 Rearrange these sentences into a correct order.

The electric bell
(a) This switches the electromagnet off.
(b) This pulls the hammer sideways to strike the gong.
(c) The electromagnet becomes magnetised.
(d) The sequence then starts again.
(e) The flow of electricity stops.
(f) When the bell is switched on, electricity flows round the circuit.
(g) This sideways movement separates the contacts.
(h) The contacts meet again and the electricity flows.
(i) The hammer springs back to its starting place.

9 Most resistors have coloured stripes painted on the side. These stripes tell you the value of the resistor in ohms. There are four stripes at one end. The first two stripes tell you the first and second numbers. The third stripe tells you how many noughts to add. The fourth stripe tells you how accurate the value of the resistor is.

This is the colour code:

Black	0
Brown	1
Red	2
Orange	3
Yellow	4
Green	5
Blue	6
Purple	7
Grey	8
White	9

Copy this table. Fill in the gaps.

Colours			value /ohms
yellow	purple	black	47
brown	black	brown	100
brown	black	–	1000
red	red	red	–
–	–	–	84
–	–	–	47 000
–	–	–	100 000

5 HEAT AND TEMPERATURE

HEAT'S A MYSTERY!

You know what it feels like on a very hot summer's day – and on a chilly winter one, too! But what is heat? Where is it kept? How does it get from one place to another?

To answer these questions, and understand what heat is, you must first know about atoms and molecules and how they move.

Heat and kinetic energy

All things are made from very very tiny particles called **atoms**. Most of these atoms are joined together in little groups called **molecules**.

Atoms and molecules never stand still. They are moving all the time. You can't see the movement of one atom or molecule, though. This is because they are so small. Like anything that is moving, atoms and molecules have energy (see page 96). We call it **kinetic energy**. If you give heat energy to water it increases the kinetic energy of its molecules. In other words, the molecules move faster.

The water in this kettle is getting heat energy

Ice, water and steam

In a lump of ice, such as an ice lolly, the water molecules are very sluggish. They don't move far. They stay in one place and vibrate backwards and forwards. This is the same in all solids. It is the reason why they keep their shapes.

Molecules in liquid water have much more kinetic energy. They have enough to move around and take up the shape of the container they are in. Bath water is warmer than ice because its molecules are moving quicker. So they have more heat energy. When water boils, its molecules have enough energy to leave the liquid and become a gas. We call this gas steam, of course! Gas released from a container will have enough kinetic energy to move all around the room.

So, to go from ice to water to steam you have to increase the kinetic energy of water molecules. This takes heat energy.

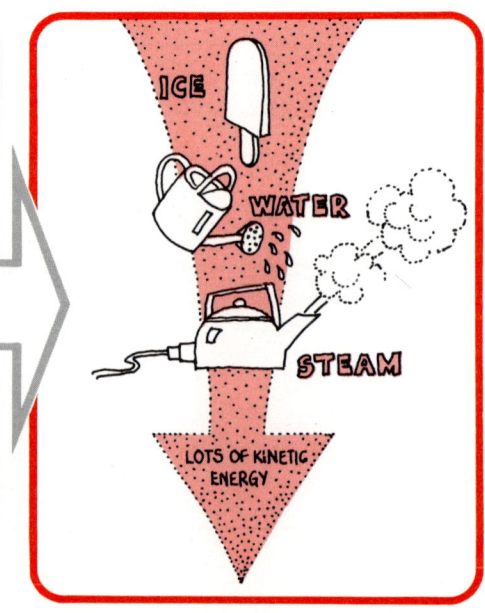

TEMPERATURE AND HEAT ENERGY

When you give water heat energy its temperature increases. Temperature is a measure of how *hot* things are.

A burning match has a higher temperature than a hot bath, but it does not have as much heat energy.

If you put the match into the water, it will go out. Almost immediately the match will have the same temperature as the bath water. The match will have cooled down a lot. But the bath will have warmed up only a tiny amount. The heat energy has moved from the hot match to the cooler bath water even though the bath has more heat energy. It is the difference in *temperature* between two objects that controls the movement of heat energy between them.

The temperature scale

We usually measure temperature on the **Celsius scale**. This scale is more commonly known as the **centigrade scale**. Here are some typical temperatures, measured in degrees Celsius (°C).

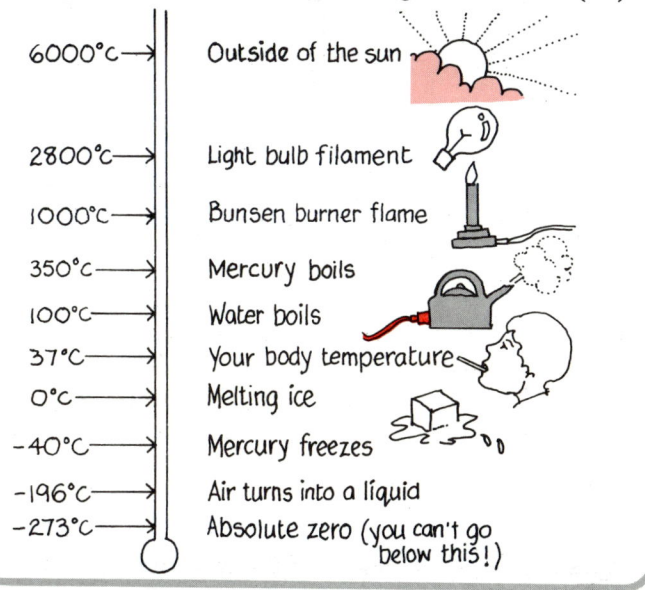

- 6000°C → Outside of the sun
- 2800°C → Light bulb filament
- 1000°C → Bunsen burner flame
- 350°C → Mercury boils
- 100°C → Water boils
- 37°C → Your body temperature
- 0°C → Melting ice
- -40°C → Mercury freezes
- -196°C → Air turns into a liquid
- -273°C → Absolute zero (you can't go below this!)

Thermometers

You can measure temperature with a **thermometer**. The most common type of thermometer has mercury inside a very narrow glass tube.

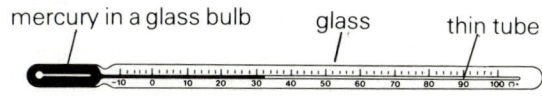

When the bulb is put into something hotter, the mercury expands up the tube.

Clinical thermometers

The best way to measure your body temperature is with a **clinical thermometer**. This is a special kind of mercury thermometer. It is made so that the mercury does not go back into the bulb when you take the thermometer out of your mouth.

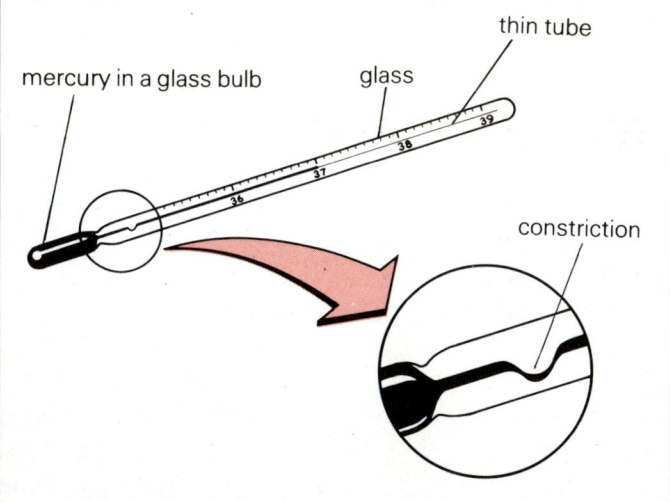

The constriction stops the mercury going back into the bulb when you take it out of your mouth. So, the thermometer continues to read your temperature even though it's cooling down.

Thermocouples

This is an industrial thermometer, called a **thermocouple**. It is made from wires of two different metals joined together at one end. When this junction is heated, a small current of electricity is made.

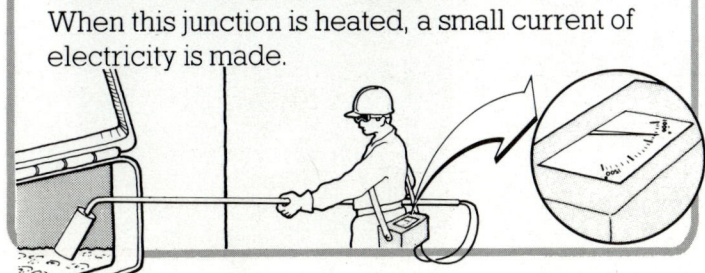

79

CONDUCTION

The handle of a metal spoon left in a cup of coffee soon becomes hot.

When the spoon is first put in, its atoms start to vibrate more. The cold end becomes hotter as the vibrations are passed along. This passing along of vibrations is called **conduction**. It stops when the ends of the spoon are at the same temperature.

Conductors and insulators

All metals are good conductors of heat. But there are plenty of materials that are poor conductors. They are called insulators. In this picture the marbles have been stuck to the ends of the rods with wax. Boiling water is then poured into the tank. The marbles on the metal rods drop off first.

The other two marbles stay firmly attached as very little heat moves up the wood and glass rods. They don't conduct heat very well. Glass and wood are insulators.

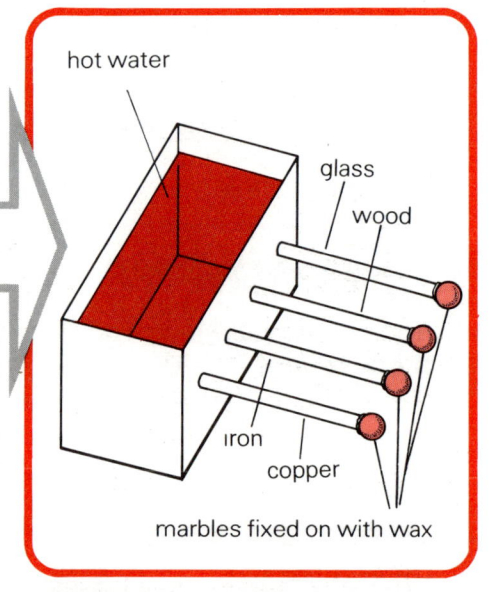

hot water
glass
wood
iron
copper
marbles fixed on with wax

Insulating with air

Air is a very good insulator. Birds and mammals make use of this to keep warm.

The owl in the picture has fluffed up its feathers. This traps air in between the feathers and its body. The air will not conduct its body heat away. So the owl stays warm. Mammals do the same thing with their fur. Tiny muscles in their skin pull each hair upright. Air gets trapped between the hairs and forms an insulating layer. The pull of these muscles causes 'goose-pimples' in human beings.

CONVECTION CURRENTS

A smokey garden bonfire shows you that warmed air rises. A similar thing happens with warmed water. It rises when it is heated. Water and air are poor conductors of heat. But they can take heat from one place to another if they are able to move. We call this **convection**. The movements of water and air are called **convection currents**.

The warm smoke rises

Air

A fire in a room provides a convection current of warm air. Look at this picture.

IT WOULD BE WARMER TO HAVE THE ARMCHAIR ON THE CEILING...!

Water

A small bottle of warm coloured water has been lowered into this tank of cold clear water. The warm water rises out of the bottle. It goes to the surface where it cools. It then sinks back to the bottom of the tank. The colour shows the convection current.

You can make your own convection current in the Activity on page 90.

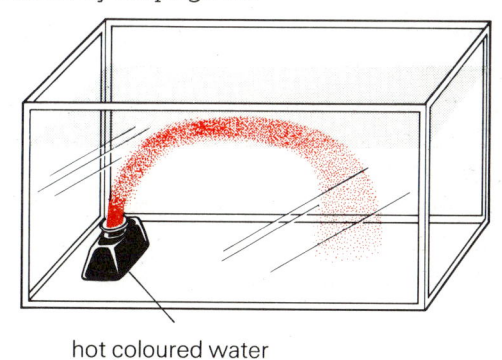

hot coloured water

Flying on warm air

On warm summer evenings it is sometimes possible to see clouds of insects above churches or trees. The clouds look like smoke, but they are made of thousands of tiny flying insects. The insects are flying in a convection current of warm air.

Some birds make use of convection currents of air which run up buildings or cliffs. Hang-glider pilots watch out for birds riding on convection currents. (Why do you think they do this?)

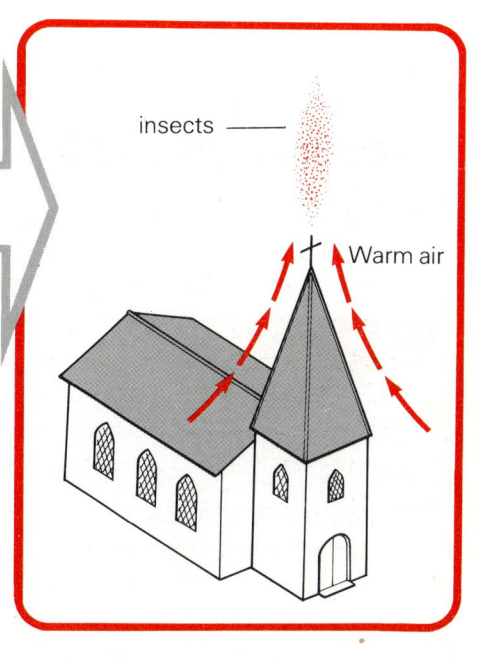

insects

Warm air

INSULATION IN THE HOME

Heat energy can be kept inside your house if the house is properly insulated. Many different insulating materials can be used. But the real insulator in all these materials is air. The different materials simply trap the air in some way and stop it moving. Proper house insulation saves energy resources such as coal and oil. It also keeps heating bills as low as possible.

Loft insulation

You can use a fibreglass mat to insulate your loft. It traps air between the fibres. At least 10 cm of fibreglass is needed. Most of the heat loss in a house *without* insulation is through the roof.

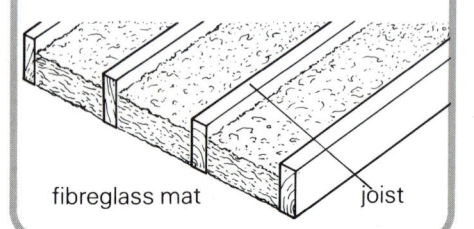

fibreglass mat   joist

Double glazing

glass

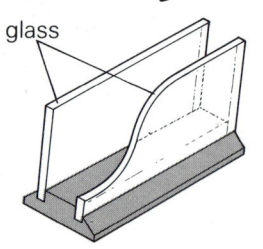

Double glazing is a sandwich of glass and air. The air gap prevents heat loss.

Save it!

Carpets

carpet   foam underlay

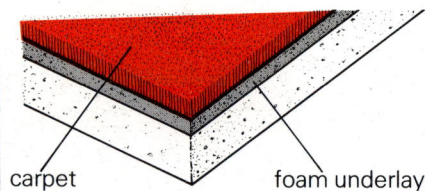

Thick carpets and foam rubber underlay will trap lots of air.

Lagged water tank

The blanket around the hot water tank in the bathroom keeps the water hot, and saves a lot of money on water heating.

Cavity wall insulation

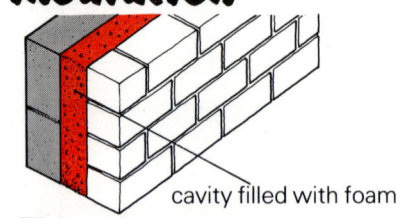

cavity filled with foam

The cavity wall space can be filled with foam containing millions of air bubbles.

USING CONVECTION

Your hot water system

In many homes, the hot water system works by convection. Water is heated in the boiler. It rises by convection to the hot water tank. Its place is taken by cold water from the bottom of the hot water tank. This continues until all of the water reaches the right temperature. The thermostat then switches off the boiler.

Should the thermostat break and the boiler *not* switch off then the system is still safe. Unwanted hot water is 'dumped' into the cold water tank through the expansion pipe.

When hot water is used, its place is taken in the system by cold water from the storage tank. The storage tank is then topped up automatically from the mains water supply.

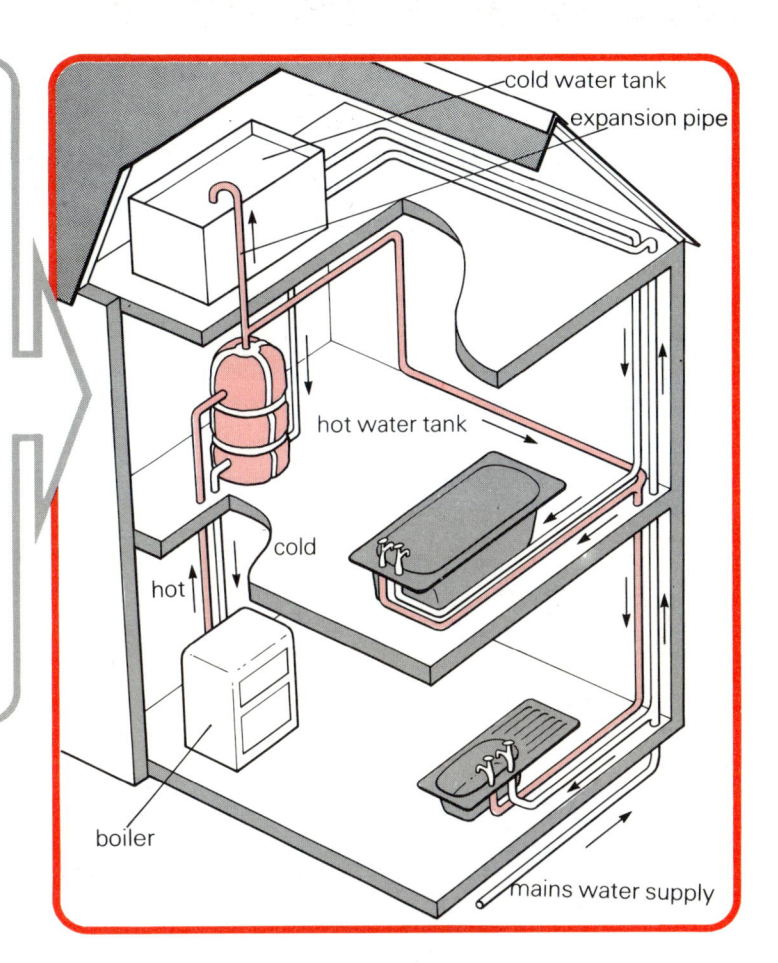

Hot air balloons

Hot air balloons work because hot air rises above cold air. The gas burners heat the air which is trapped in the balloon. The hot air rises and the balloon goes up taking the balloonists with it. As the air cools, the balloon stops rising and drifts downwards. To gain height, the balloonists turn on the burner again. Have you ever seen a real hot air balloon?

The burner heats the air in the balloon

MELTING AND BOILING

Melting

As you discovered on page 78, solids turn into liquids when their molecules break free from their places and start to move about. This process is called **melting**. A solid needs energy to do this.

The energy needed to make a solid melt is called **latent heat**. The latent heat for ice can be measured like this.

The **joulemeter** measures the number of **joules** (J) of energy given by the power supply to melt the ice. It takes about 34 000 J to melt 100 g of ice.

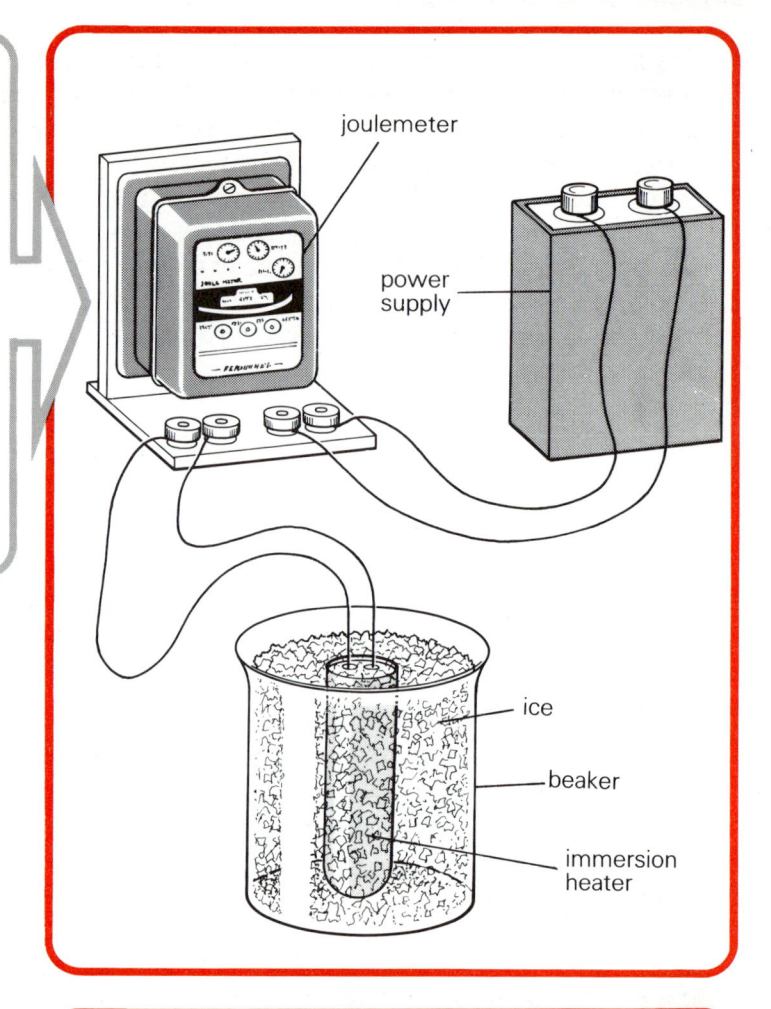

Melting point

A pure substance melts at a particular temperature. This is called its **melting point**. For example, the melting point of pure ice is 0°C. Melting happens when a solid is heated after its temperature has reached the **melting point**. Freezing is the opposite thing! Freezing happens when a liquid is cooled after its temperature has reached the **freezing point**.

The freezing point and the melting point are exactly the same.

Adding salt to water lowers the water's melting point. This is why salt is put on icy roads. The salty ice melts because its melting point is lower than the temperature of the road.

Other impurities can be added to water to lower its melting point. Motorists put **anti-freeze** into the water-cooling systems of their cars. The anti-freeze and water mixture has a very low melting point. So, it will not freeze, even in the worst winter!

Salting the roads in cold weather

Anti-freeze stops the water in the cooling system freezing

EVAPORATION

The molecules in a liquid are able to move about within it. They need energy though, to escape from the liquid, and become gas. Changing from liquid to gas is called **evaporation**. The energy needed is called **latent heat**.

Puddles of rain disappear, even on cold days. This is because the water molecules receive enough energy to evaporate.

A few drops of methylated spirit, or nail-varnish remover, on the palm of your hand will quickly evaporate. The energy needed to do this is taken from your hand. So, your hand feels cold when this happens!

This puddle will soon evaporate

Keeping cool

Your body's cooling system depends upon evaporation. When you get hot, you sweat. The sweat evaporates into the air, using latent heat energy that it takes from you. So, you get cooler!

Dogs cannot sweat efficiently because of their hairy coats. They lose heat by panting and sticking out their tongues. Evaporation from their mouths cools them down.

Boiling point

Evaporation normally takes place at the surface of the liquid. **Boiling** is evaporation *anywhere* in the liquid. It happens at a special temperature called the **boiling point**. The boiling point for pure water at normal air pressure is 100°C.

The boiling point of water is higher than 100°C if the pressure of the air is greater than normal. The boiling point of water is lower than 100°C if the pressure is lower than normal. You can't make a decent cup of tea on the top of Mount Everest because water boils at 70°C!

Fridges and freezers

Fridges work because evaporation uses energy. A special liquid, called **freon**, is pumped round inside pipes surrounding the freezing compartment. The freon evaporates using heat energy from the food. So, the food gets colder.

The pump compresses the freon vapour and turns it back into liquid. The freon gives up latent heat as it becomes a liquid again. The heat goes out into the kitchen through the cooling fins.

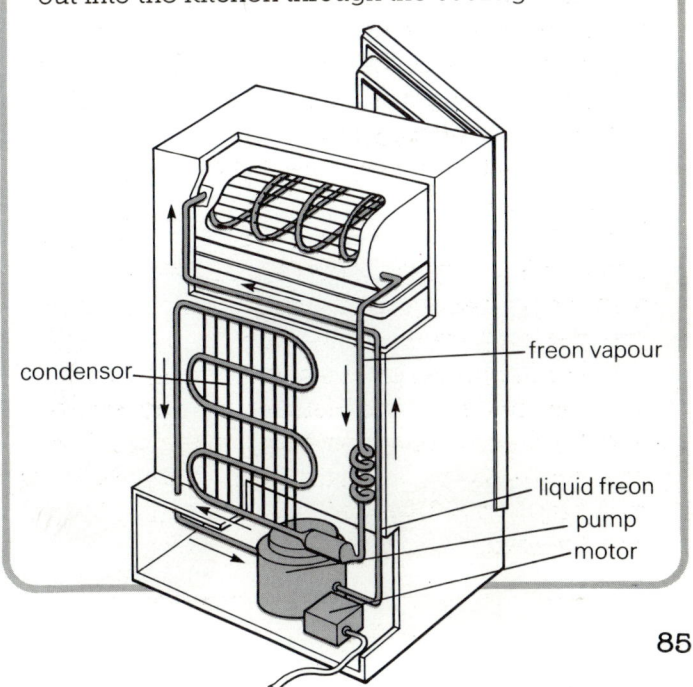

- condensor
- freon vapour
- liquid freon
- pump
- motor

85

EXPANSION — 1 SOLIDS

How do solids expand?

Most solids get bigger when they are heated. The molecules vibrate further as the temperature goes up. They need more room to do this, which causes **expansion**. The expansion is very small, but its effects can be very important. They may be disastrous, but they can be very useful! When a solid cools, its molecules vibrate less and the solid gets smaller. We say that it **contracts**.

Steel bridges

This bridge is made of steel. It expands as it gets warmer. The length of the bridge could increase by several centimetres on a hot day. If this were not allowed for in the design of the bridge, the huge expansion forces would buckle it.

The bridge is resting on rollers so that it can expand and contract. The gap is small in summer and large in winter. It is usually covered with a narrow steel plate.

Railway lines

Railway lines are joined so that they overlap like this. A continuous length of railway line would buckle on a hot day.

Opening lids

If a metal screw-top on a glass bottle or jar is too tight to unscrew, you can free it like this. Run the hot tap over the top for a few minutes. Heat gets to the top and it expands. Less heat reaches the glass so it hardly changes. This loosens the top.

Bimetal strips

A **bimetal strip** is made of two different metals joined together. One of the metals expands a lot when heated. The other hardly expands at all. When the bimetal strip is heated it bends. The metal with the greater expansion is on the outside of the bend.

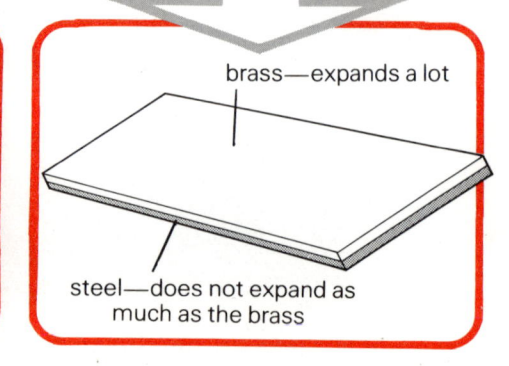

brass—expands a lot

steel—does not expand as much as the brass

EXPANSION — 2 LIQUIDS

How do liquids expand?

When liquids are heated they expand as solids do. Liquids expand *more* than solids, though. Different liquids expand by different amounts. Look at the picture. When this trough is filled with hot water, the liquids expand up the tubes. They reach different levels. Expanding liquids can be quite useful (look at page 79).

Water - a special case

Water is different from other liquids. It *contracts* if it is heated from 0°C to 4°C, but it expands if heated above 4°C. This means that water has its smallest volume at 4°C.

As a pond gets colder and colder in winter, water at 4°C sinks to the bottom. Water that is colder than 4°C goes to the top. So, the pond freezes over from the top down leaving water underneath the ice.

Gases

Gases expand when heated even more than liquids.

If you do this to a flask of air, the air expands. You will see the bubbles of air coming out of the tube. When you take your hands away, the flask cools and the air contracts. Some of the water will be sucked up the tube.

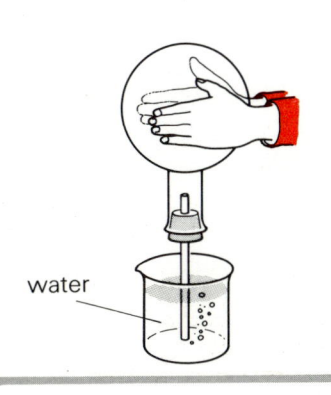

Absolute zero

Gases contract when they are cooled. If you could take a flask of air and measure its volume as it was cooled, you might get a graph like this.

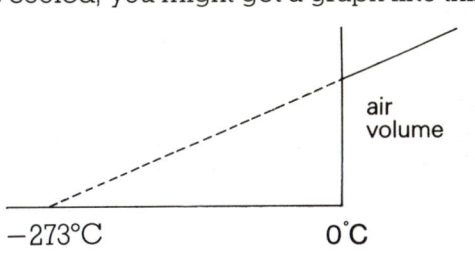

The dotted line shows the results that you might expect if you carried on cooling the flask. This line predicts that, if the air stayed as a gas and did not turn into a liquid, *it would have no volume at all at −273°C.*

This is the **absolute zero of temperature**. This temperature is the most sensible point to use to start a temperature scale. It is the lowest possible temperature.

The **Kelvin scale** does just that. −273°C is called 0 K (zero kelvin). Each kelvin is equal to 1°C.

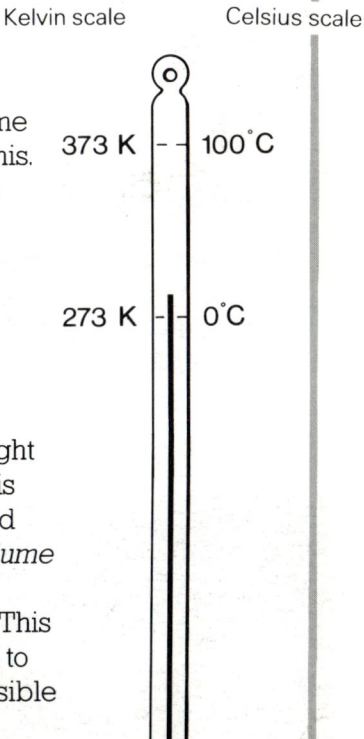

Kelvin scale / Celsius scale

373 K — 100°C
273 K — 0°C
0 K — −273°C

WHAT IS THE MELTING POINT OF WAX?

Wax has a melting point between 0°C and 100°C. By taking the temperature of wax as it cools from 100°C you can find out exactly what the melting point is.

You will need

- boiling tube containing wax
- thermometer
- beaker
- water
- tripod
- Bunsen burner
- stop-watch

Copy this table

Time / minutes	Temperature / °C

1 Put the boiling tube containing wax in the beaker. Clamp the boiling tube for safety. Pour water into the beaker until it's just above the level of the wax.

2 Put a thermometer in the tube. Heat the water until it boils. Let it boil until all of the wax has melted.

3 Take the boiling tube out of the beaker. The wax will begin to cool. Take its temperature every half minute. Record your results in the table.

4 Stop taking readings once the temperature has dropped to about 30°C. Plot your results on a graph like this.

Answer these questions

1. Why is it important to keep on heating until all of the wax has melted?
2. What do you think the melting point of wax is?
3. How does the freezing show up on your graph?
4. Is the wax cooling down at the melting point?
5. Is the wax losing energy at the melting point?
6. What do you think the shape of the graph would be if you did the experiment while heating up the wax?

MAKING A FIRE ALARM

Metals expand when they are heated. Different metals expand by different amounts. A bimetal strip is made from thin strips of two metals squashed together. When it is heated, the metals expand by different amounts, so the strip bends. You can use this idea to make a fire alarm.

You will need
- bimetal strip
- leads with crocodile clips
- power pack
- electric bell (or ray box)
- Bunsen burner
- clamp and stand

Set up this circuit
   Turn on the power pack to 12 V.
   Light the Bunsen burner and turn it to a roaring flame.

Answer these questions

1 What happens as the bimetallic strip warms up?
2 What would be the effect of making the gap between the bimetal strip and the plug smaller? What would happen if you made the gap bigger?

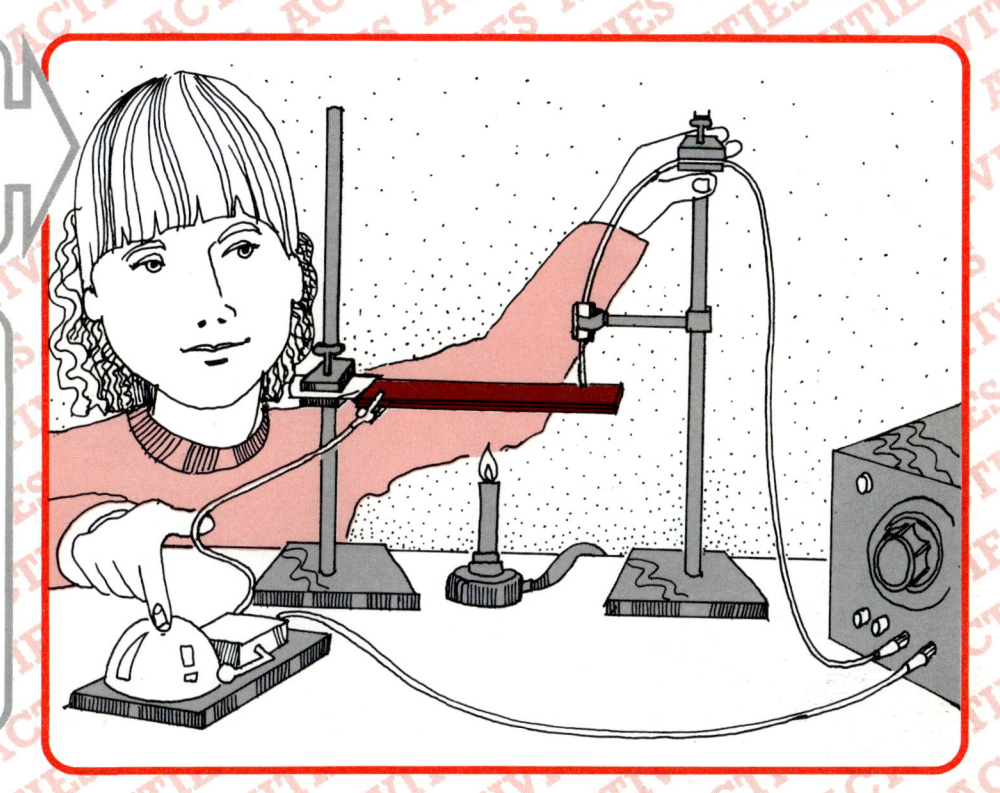

How do car indicator lights work?

The indicator lights on a car use a bimetal strip. The strip has a heater wrapped around it. When the indicator is switched on, the bulb lights up and the heater is turned on. The heater warms up very quickly. The heat bends the bimetal strip upwards and the contact is broken. This switches the bulb and the heater off.

As the bimetal strip cools, it goes back to its original shape. Contact is made again. The bulb lights up. The heater is turned on and the whole cycle starts again. The bulb keeps flashing until the indicator is switched off.

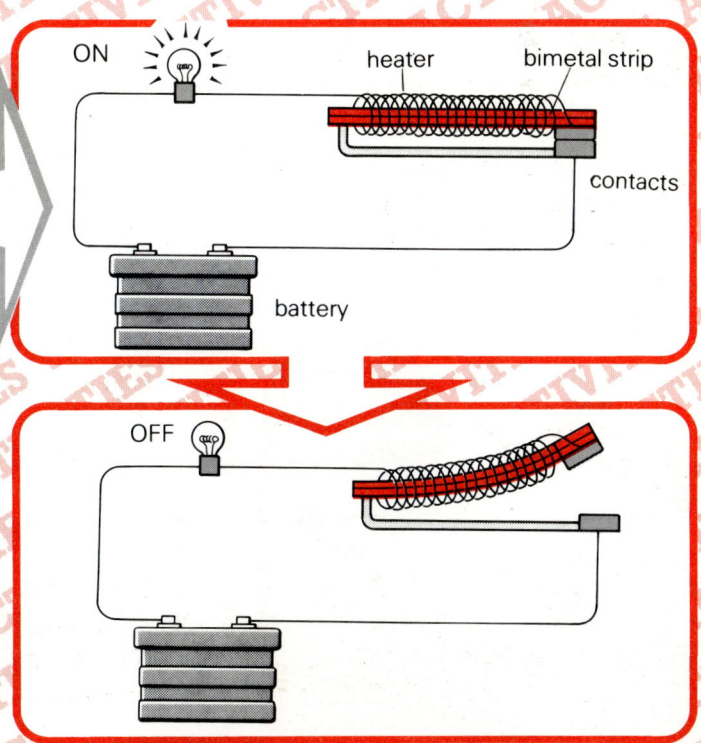

FINDING OUT ABOUT CONVECTION CURRENTS

You find convection currents in liquids and gases when you heat them. The hot part of the liquid (or gas) rises upwards to be replaced by cold liquid (or gas). This cold part is then heated in turn. It rises and is replaced by cold. These movements are **convection currents**. In this Activity you can investigate some of these convection currents.

You will need

- potassium permanganate crystals
- sawdust
- glass tubing
- water
- 250 cm³ beaker
- aluminium foil
- scissors
- thread
- Bunsen burner
- tripod
- lemonade or squash bottle

1. Pour about 200 cm³ of water into the beaker. Put the tube into the water. Drop a crystal of potassium permanganate down the tube.

2. Put your finger over the end of the tube. Carefully lift it out. Let the water in the tube go down the sink.

3. Gently warm the beaker underneath the crystal. Use the smallest yellow flame you can. Describe what you see in the beaker. Explain why it is happening.

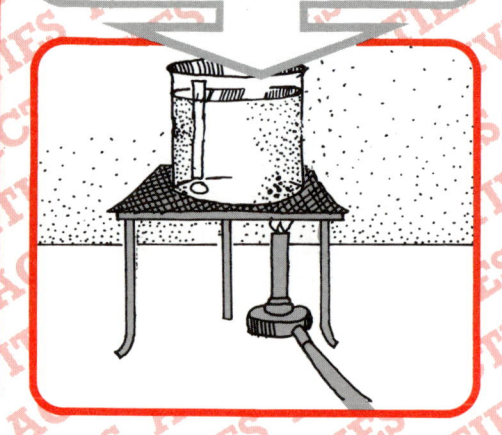

4. Put a handful of sawdust into a lemonade bottle. Fill it with water and shake it. Stand it on a radiator. Watch it to see what happens.

5. Mark out these shapes on a piece of aluminium foil. Cut them out.

6. Hang them from thread over the small Bunsen flame. Try other shapes.

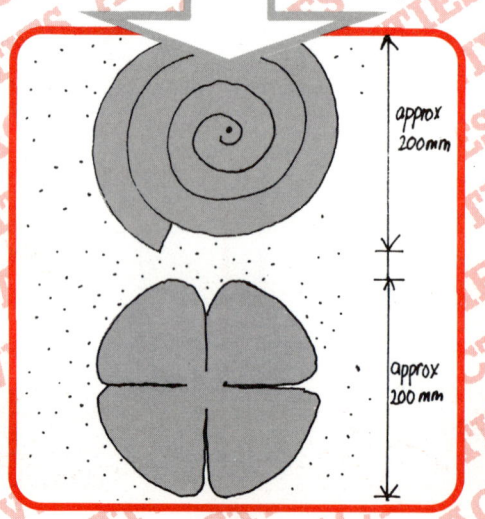

HOW GOOD IS A VACUUM FLASK?

There are two kinds of flasks you can use to keep drinks warm. There is the older 'breakable' vacuum kind and the newer 'unbreakable' sort. This is how the two types look inside:

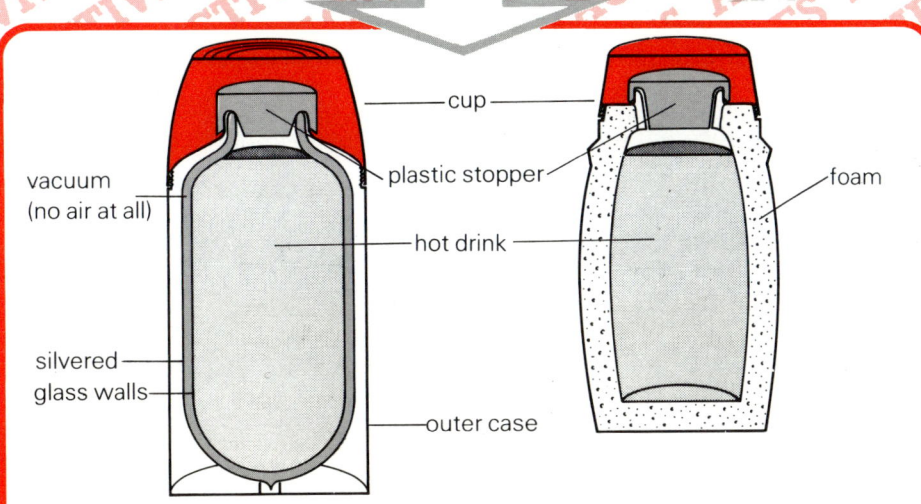

You will need

- 250 cm³ beaker
- electric kettle (or Bunsen burner)
- long thermometer
- vacuum flask
- unbreakable flask
- oven gloves

Labels on breakable flask: cup, plastic stopper, hot drink, vacuum (no air at all), silvered glass walls, outer case

Labels on unbreakable flask: foam

1 There is nothing in the walls of the flask to conduct or convect the heat. Even the stopper is insulated. The silvered walls of the flask stop heat escaping by radiation.

2 The air trapped in the foam is a good insulator. It cannot move to take heat energy away by convection. It doesn't conduct very much heat energy away either.

Copy this table

Time / minutes	Temperature / °C		
	Vacuum flask	Unbreakable flask	Beaker

1 Boil a kettle of water. Pour 200 cm³ of the boiling water into the beaker. *Using oven gloves*, quickly pour it into the vacuum flask. Do the same with the other flask. Finally, leave a third 200 cm³ of boiling water in the beaker.

2 Take the temperatures of the water in both flasks and the beaker. Put your results in the table. Repeat this every 10 minutes for as long as you can.

Answer these questions

1. Draw a graph of your results.
2. Which flask kept the water warmer?
3. Why was it useful to include the beaker in this Activity?
4. Which flask is better at keeping in heat energy?
5. You might take the results of this Activity into consideration when you buy a new flask.
   What other factors would you have to think about?
6. Which flask would keep cold drinks cooler on a hot summer's day? How could you alter this Activity to find out?

91

EXERCISES

1. What is the difference between heat and temperature?

2. Mercury thermometers don't just happen to be the shape they are! They are designed like that so they work as well as they possibly can. Draw a diagram of a mercury thermometer. Label the diagram. Describe the parts which help it to work well.

3. A man on television today said that it was so cold on the planet Uranus that the temperature was −300°C. Why was he wrong?

4. Clinical thermometers are never sterilised in boiling water. Why not? What *is* used?

5. Imagine you are a molecule in an ice cube. Describe what would happen to you when the ice-cube you are in is taken from the coldest part of the freezer, placed in a pan and heated until the pan boils dry.

6. Decide whether these statements are true or false. If they are false explain why.
   (a) Motorists put anti-freeze into their car engines to raise the melting point of the water in the cooling system.
   (b) Water shrinks when it is heated from 0°C to 4°C.
   (c) A bimetal strip bends when heated because one of the two metals expands more than the other.
   (d) Air is a very good conductor of heat.
   (e) Molecules in a gas move faster than in a solid.

7. (a) Furniture and houses often creak as the house cools down at the end of a very hot day. Explain why this happens.
   (b) Why do you feel chilly when you get out of the swimming pool? Why do you feel warmer when you are dry?
   (c) Is it possible to keep the kitchen cool on a hot day by opening the fridge door?
   (d) Why should you never throw an aerosol can on an open fire?
   (e) Pans are made of metal, so that heat can be conducted to the food quickly. Why do they have plastic or wooden handles?

8. Suppose that you could choose to have the best sort of day for drying your swimming towel and costume. What would it be like? Would it be sunny or cloudy? Windy or calm? Dry or humid? How would you hang them on the line so they dry quickly?

9 The following readings were recorded by a girl and a boy working on the melting point of wax Activity (page 88).

Time / minutes	Temperature / °C
0	91
2	65
4	35
6	55
8	49
10	43

Plot a graph of these results.
Answer these questions.
(a) What was the temperature of the wax after 8½ minutes?
(b) When was the temperature of the wax 65°C?
(c) What was the melting point?
(d) When was the wax cooling fastest?
(e) Was the wax cooling down between 4 minutes and 6 minutes?
(f) Was the wax losing energy between 4 minutes and 6 minutes?

10 Draw a diagram of your house. Mark on it the places where insulation is, or could be used, to save energy.
Can you think of a way of checking that your insulation is really saving energy? What measurements would you have to take? What information might you need?

11 What happens in a house hot-water system when someone draws off a nice hot bath? Use the diagram on page 83 to help you explain.

12 Design and draw a frost alarm, using the idea of the fire alarm in the Activity on page 89. How could you be sure that it rang out when the air temperature got down to freezing (0°C)? Who would find one of these frost alarms useful, and why?

13 Copy out this passage and fill in the missing words:

"When a liquid turns into a .... , we say it evaporates. When this happens anywhere in the liquid, the liquid is .... This only happens at a special temperature, called the .... Water boils at .... in normal air pressure. If the air pressure is lower, for example at the top of a .... then the boiling point will be .... too."

93

6 ATOMS AND MOLECULES

KINETIC THEORY

There is a lot of evidence to tell us that all things are made of particles called **molecules** and that these particles are moving all the time. There is also a lot of evidence to tell us that the *amount* these molecules move depends on their temperature. Together these ideas are known as the **kinetic theory**. In this chapter you will be able to find out about some of the evidence that supports this theory.

Solids

The atoms or molecules in a solid are held together by strong forces of attraction. They *do* move a little, but not enough to leave their position in the solid. They can only vibrate backwards and forwards on the spot. A solid always has a fixed shape because the particles in it can't move far. The atoms or molecules do not have enough energy to break free from the rest of the solid. Look at the top picture This is a model of a crystal. A pattern like this is called a lattice. The atoms inside it fit together to form a crystal. You can see that crystals have lattices from their regular shape. More evidence for the lattice structure of crystals comes from the splitting of crystals into regularly shaped pieces. You can do this by striking a crystal with a razor blade. You get the cleanest cut if the blade is held at the same angle as the crystal surface. The razor separates layers of the lattice.

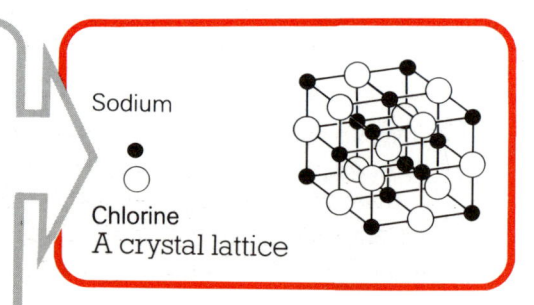

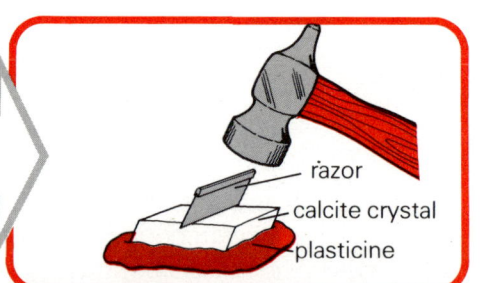

Sodium
Chlorine
A crystal lattice

razor
calcite crystal
plasticine

Liquids

When a solid is warmed up, its molecules move faster (remember Chapter 5?). They still stay in the same place, but they vibrate with more energy. As the heating continues the temperature of the solid increases until its molecules have enough energy to leave the lattice. When this happens, we say that the solid has melted and turned into a liquid.

Molecules in a liquid have too much energy to stay in one place. This means that a liquid moves to take up the shape of the container holding it.

Gases

Molecules in a liquid move faster as the temperature increases. Eventually, they gain enough energy to break free and leave the liquid. When this happens we say that the liquid is evaporating to become a gas. Gas molecules move at high speed. They completely fill their container. So air in the room you are in goes into every part of it.

DIFFUSION

What is it?

Molecules in a gas move around so quickly that they are always bumping into each other. If possible they will always move from a place where they are crowded together to a less crowded place. This movement is called **diffusion**.

Smells are a good example of diffusion. When you smell something, molecules from the thing that is smelly have diffused through the air and gone into your nose. Have you noticed how quickly smells move around a room? Gas molecules are fast movers!

The experiment on the right shows diffusion very clearly.

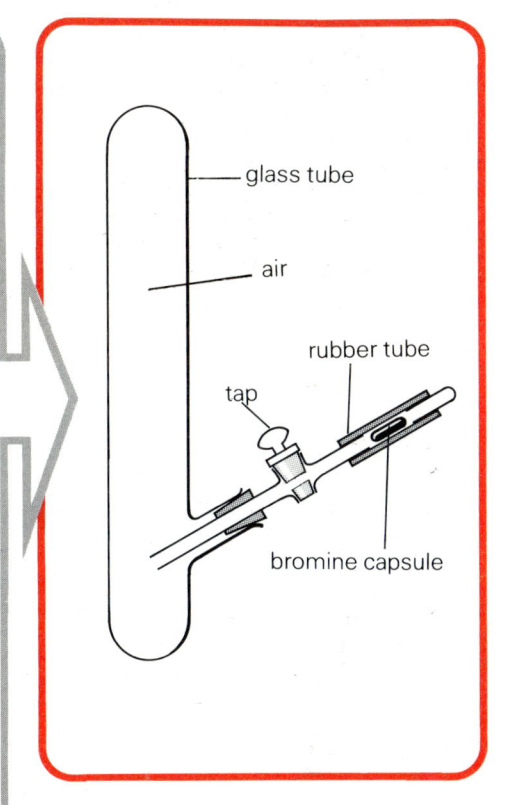

Bromine is a poisonous gas. It is ideal for this experiment, though, because it is dark brown and easy to see. The bromine capsule is broken with pliers. The tap is turned on. The gas slowly works its way up the tube as diffusion takes place.

If there was no air in the tube the bromine gas would fill it almost straight away. Can you think why?

Diffusion also happens in liquids, but it's much slower than in gases. This is because molecules in liquids move more slowly and are more crowded together. So, when they collide they don't get knocked far. Diffusion happens in solids, but so slowly that you can't easily tell.

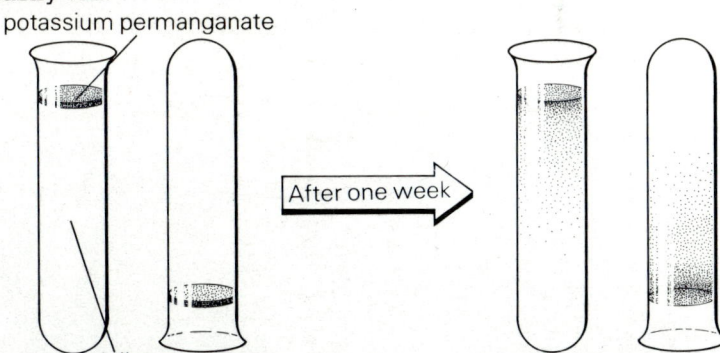

This experiment shows diffusion in liquids.

The two test tubes are filled with jelly. The jelly stops convection currents spoiling the experiment.

The potassium permanganate has dissolved in the water at the jelly surface. It has diffused through the water. Why have one of the tubes upside down?

What is it?

Have you ever noticed that drips of water from a tap are almost perfect spheres?

Surface tension in action

Have you ever seen an insect called a pond-skater walking on water?

Surface tension in action

Have you ever made a needle float on water? (Float the needle on a piece of tissue paper first. When the tissue paper soaks up water and sinks, the needle will float by itself.)

Surface tension in action

Why does it happen?

Surface tension makes it seem as if there is a skin on the surface of the water. There is no skin in fact, but the water behaves as though there is. Surface tension happens because the water molecules are attracted to each other.

water molecule
attractive forces

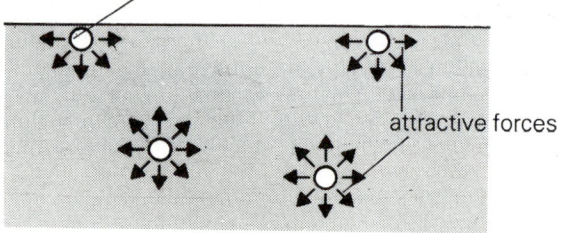

A water molecule in the middle of the water has a balance of attractive forces all around. But a molecule at the surface only has forces pulling inwards. This inward pull is the surface tension.

The inward pull on a drop of water tends to make it spherical.

A water droplet

Soaps and detergents

Washing powders lower surface tension in the washing water. This then lets the water get between the fibres of the clothes to dissolve out the dirt.

It is difficult to blow bubbles with pure water. This is because the surface tension is so strong. Any bubbles you make burst very quickly. Soap solution, however, has a lower surface tension. It is low enough for the bubbles to last.

Soap lowers the surface tension

96

The surfaces of liquids

If you look at the surface of water in a glass, the water seems to have crept up the sides a little. The curved surface is called the **meniscus**. It is this shape because of the attraction between molecules.

The water molecules are attracted to each other but they are more attracted to the glass molecules. So, the water tends to cling on to the glass making the meniscus curve up at the edges. The attraction of glass for water is the reason why water wets glass. It also explains why, when you empty a glass of water, there are always a few drops left behind.

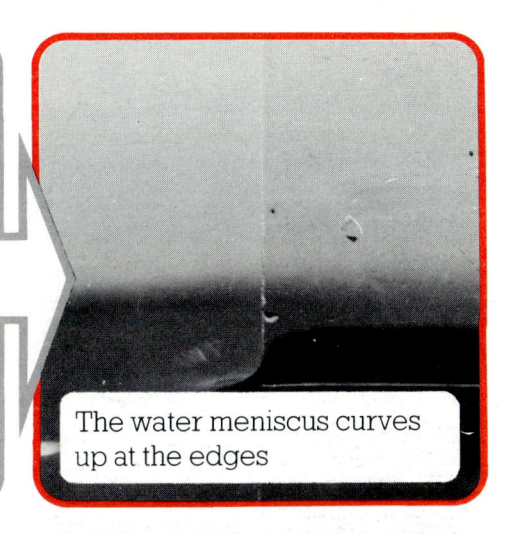

The water meniscus curves up at the edges

Funny liquids

With some liquids, like mercury, the opposite is true. Their molecules are more attracted to each other than they are to the glass. The mercury meniscus curves down at the edges. It is the opposite shape to a water meniscus. If a beaker of mercury is emptied, it will be completely dry.

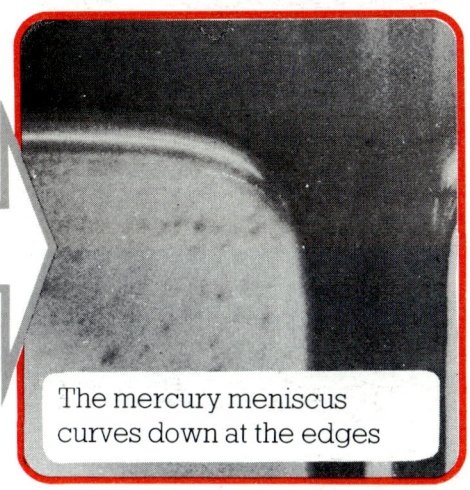

The mercury meniscus curves down at the edges

Capillary action

The attraction of glass for water is strong enough to pull water up very narrow glass tubes, called capillary tubes. This is known as **capillary action**.

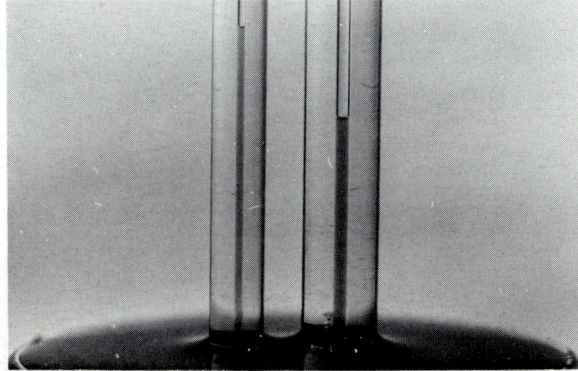

The narrower the tube, the further the water is pulled up.

Capillary action is very useful. When you dry yourself with a towel, the water is soaked up by capillary action. Paper tissues are absorbent because of it.

Capillary action can be a nuisance. Water will soak up through the walls of a house (rising damp) unless capillary action is prevented. A **damp-proof course** in the brick wall stops the rising damp.

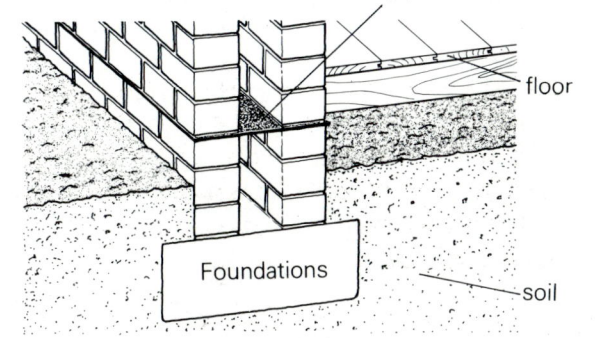

The plastic layer stops water movement. The soaking of rain water to the inside by capillary action is prevented by the air gap.

ATOMS

What's inside?

For a long time atoms were thought to be the smallest particles that things are made of. We now realise that even atoms are made up from smaller particles. These are **electrons**, **protons** and **neutrons**. We have found this out from several experiments. Some of the experiments are described on this page and in the rest of this chapter. By the end of the chapter you should understand how atoms are made up and how the knowledge of what they are like has been used.

Electrons

Electrons are very, very small particles. They are found in every atom. If you heat a wire made from tungsten to a very high temperature, some of its electrons gain enough energy to leave the wire. This is called **thermionic emission**. It is similar to evaporation, but evaporation happens to whole molecules.

Thermionic emission

This glass tube has been made to show thermionic emission. The screen glows when electrons hit it. The tube has no air in it because air molecules would stop electrons moving. The tungsten is heated by a low voltage power supply. A very high voltage is connected across the tube. The negative is connected to the tungsten filament (the **cathode**). The positive is connected to a metal plate (the **anode**). You cannot see the electrons going down the tube because they are too small to see. But you can tell when they are there because the screen glows.

If the anode is made negative, the screen stops glowing. Electrons are attracted to a positive anode but not to one that has been made negative. This shows that the electrons carry a negative charge. There is more about charges on page 100.

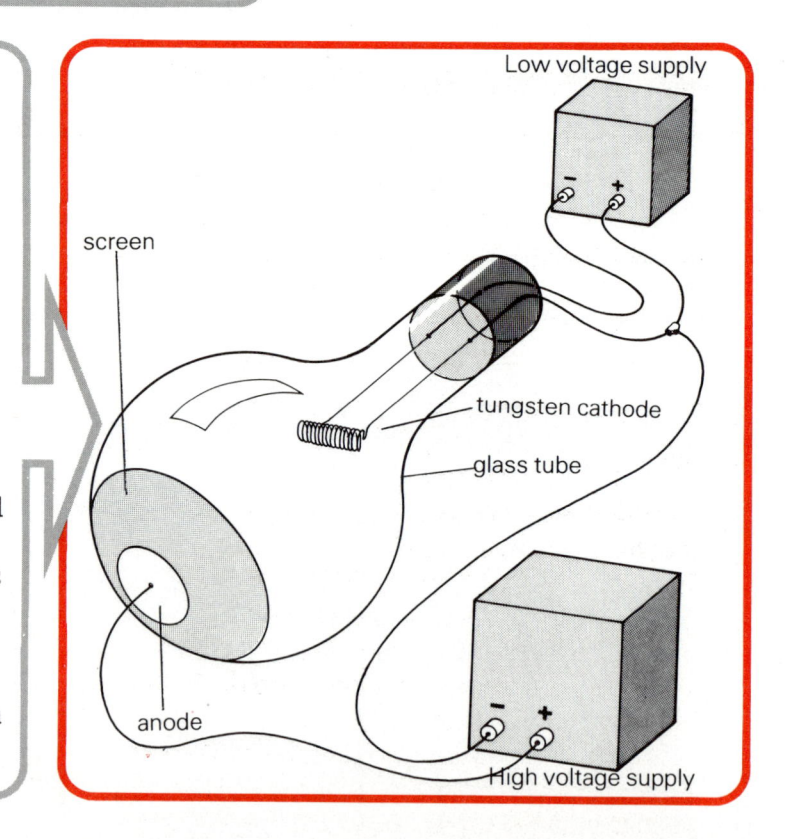

Cathode rays

The electrons released by thermionic emission can be drawn into a ray. This is done by firing them through an anode with a small hole in it. This kind of anode is called an **electron gun**.

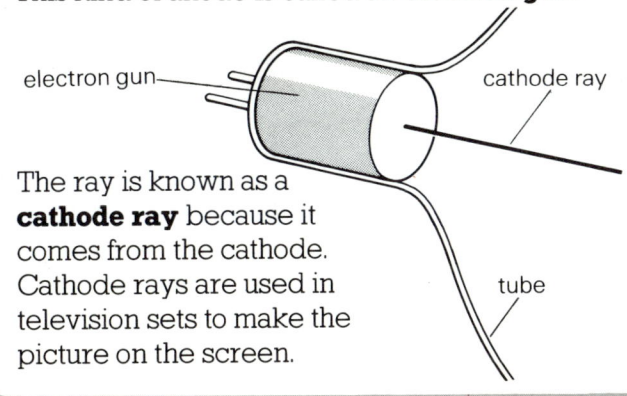

The ray is known as a **cathode ray** because it comes from the cathode. Cathode rays are used in television sets to make the picture on the screen.

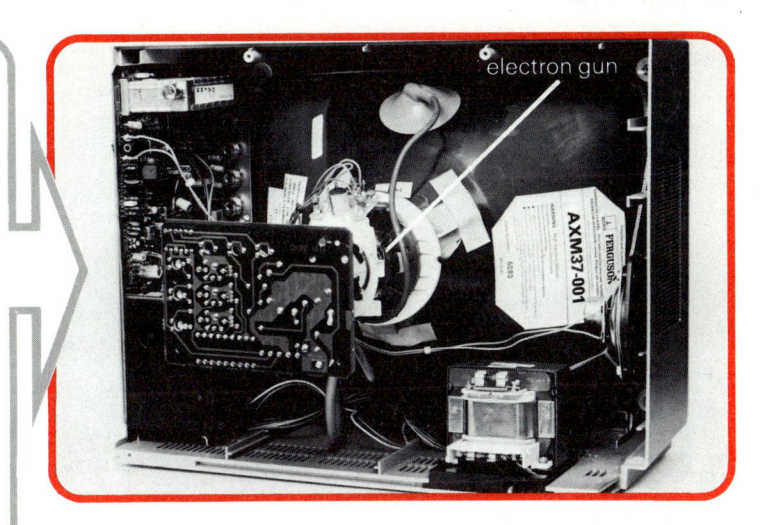

How does your TV work?

Cathode rays are made in the electron gun at the back of the cathode ray tube. They are fired at the screen at the front. You can see a target spot on the screen wherever the ray hits it. The spot is moved around the screen by two electromagnetic coils wrapped around the tube.

One coil makes the spot more slowly across the screen and quickly back again. The other moves the spot slowly down the screen and then quickly back to the top again.

While this is going on the brightness of the spot is changed by the signal picked up by your TV. The signal comes from the BBC or ITV. It has all the information needed to show you a programme. When you look at a black and white TV you actually see a picture made of all shades of grey, from white to black.

The picture is drawn by the one spot. It makes a new picture 25 times a second. Your eyes can't tell this though. All the movements appear smooth and continuous.

A colour TV uses cathode rays just like a black and white TV. The picture is built-up in the same way. You can find out how the picture appears coloured on page 113.

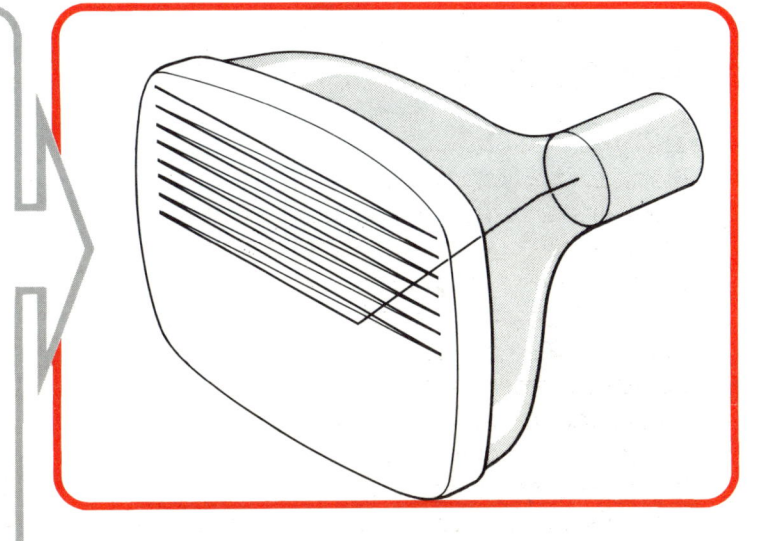

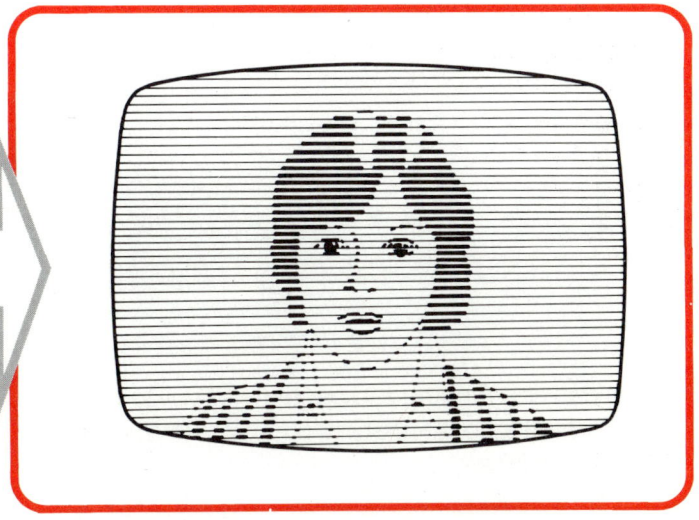

RADIOACTIVITY

What is it?

Some substances are **radioactive**. They give out energy.

The energy from a radioactive substance can be in the form of:

1 Alpha particles. These are parts of helium atoms. They carry a positive charge. They are not very penetrating. In fact, a thin sheet of paper will stop them.

2 Beta particles. These are fast-moving electrons. They penetrate further than alpha particles. It takes a thin sheet of metal to stop them.

3 Gamma rays. These are high-energy electromagnetic waves. They are the most penetrating kind of radioactivity.

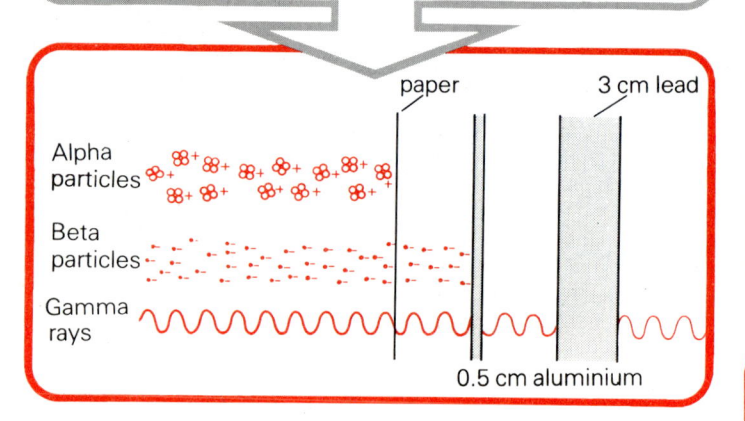

Half-life

As radioactive materials give off radiation they become less radioactive. We say they **decay**. The time taken for the radioactivity to fall to half of its original amount is called the **half-life**. This graph shows the decay or radon-220.

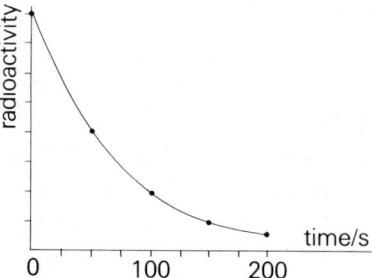

You can see that the radioactivity drops to half of its original amount in about 50 s. So, we say its half-life is 50 s.

Ionisation

Many of the effects of radioactivity – and they aren't *all* harmful – occur because it can **ionise** atoms to make **ions**. Normally, atoms have no charge on them. The positive and negative charges each atom contains are balanced. However, an electron can be knocked out of an atom. The electron takes its negative charge with it. So, the part left has a positive charge. This process is known as **ionisation**.

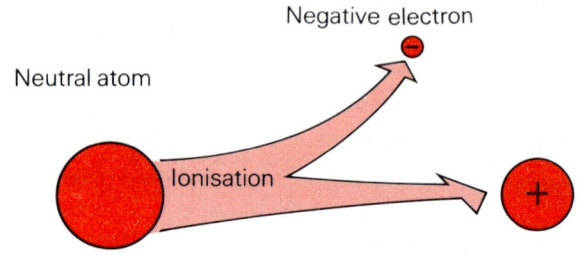

You can't see radioactivity but you can detect the ionisation it causes using a **Geiger–Muller tube**. Each ion is counted and makes a click in a loudspeaker. The tube and the counter make what is known as a **Geiger counter**. Have you ever heard the sound a Geiger counter makes?

This man is using a Geiger counter

Background count

If a Geiger counter is switched on well away from any radioactive sources, it still detects radioactivity. This is the **background count**. It is the radioactivity that we are all exposed to every day. Most of it is natural. It comes from the Sun and outer Space or the Earth itself. There is no reason to think it is harmful!

USES OF RADIOACTIVITY

The nucleus

Radioactivity was used to find out how atoms are made up. When alpha particles were fired at thin gold foil, most went straight through. But a few, about 1 in every 8000, bounced back.

This means that most of the volume of gold atoms is empty space! The material in a gold atom is concentrated in a very small, positively charged nucleus. It is the same for all atoms.

The nucleus can have two different kinds of particle. There are protons which carry a positive charge. There may also be neutrons. These carry no charge but each has the same mass as a proton.

Electrons orbit the nucleus. They carry a negative charge which is equal and opposite to the charge on the protons. In a neutral atom, positive and negative charges are balanced. There are equal numbers of protons and electrons. Electrons are very small. Each one has a mass of about 1/2000th of the mass of a proton. Most of the nucleus is empty space.
If you imagine Wembley Stadium as an atom with the electrons moving around the outside of the stadium, then the nucleus will be about the size of a hockey ball on the centre spot.

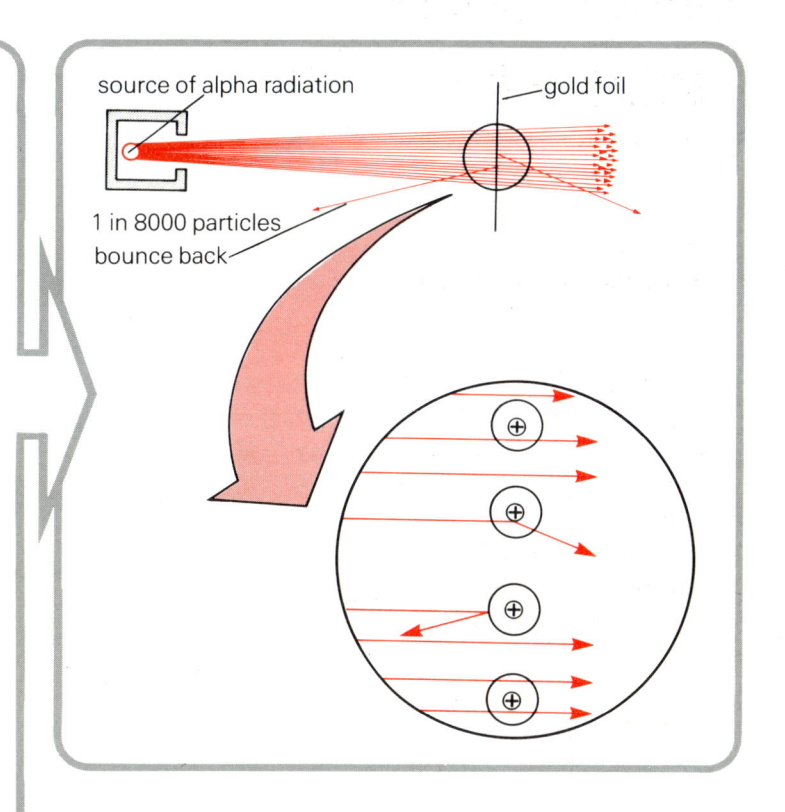

Radiation therapy

Gamma rays can be used to kill cancer cells in the human body.

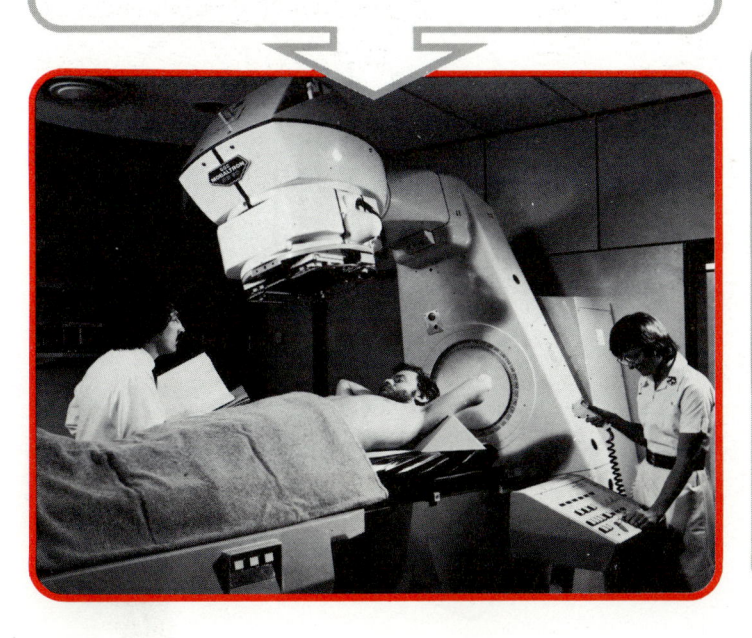

Carbon dating

Carbon, along with many other elements, is a mixture of atoms slightly different from one another. It contains ordinary non-radioactive carbon together with a very tiny amount of radioactive carbon. This is called carbon-14. The ages of fossils can be dated using the radioactivity. The proportion of the carbon-14 in animals and plants is roughly constant when they are alive. When they die the amount they contain starts to fall as the carbon-14 decays. The half-life of carbon-14 is 5600 years. So, knowing this and the activity of a sample, it is possible to work out the date when the plant or animal died.

101

NUCLEAR ENERGY

Radioactive decay gives out energy. The energy comes from the nuclei of the radioactive elements, so the energy is called **nuclear energy**.

Nuclear fission

Uranium is one of the most important elements used in the production of nuclear energy. There are two sorts: uranium-238 and uranium-235. Uranium-235 gives off nuclear energy very rapidly during a process called **nuclear fission**. When a neutron strikes a uranium-235 nucleus it splits the nucleus into two parts.

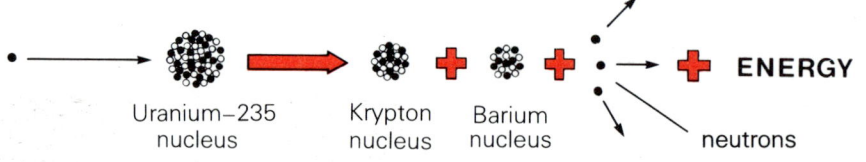

Uranium–235 nucleus → Krypton nucleus + Barium nucleus + neutrons → ENERGY

The split nucleus becomes two new atoms. A large amount of energy is given off. But, most important of all, three neutrons are released. Each one of these three can go and split another uranium-235 nucleus, each producing three more neutrons.

This is called a **chain reaction**. It occurs naturally in a sample of uranium larger than a certain mass, called the **critical mass**. In smaller pieces of uranium the neutrons escape into the air without splitting more atoms.

The first atom bomb was an uncontrolled chain reaction in uranium-235. It was started when two lumps of uranium-235 were brought together and were then together bigger than the critical mass.

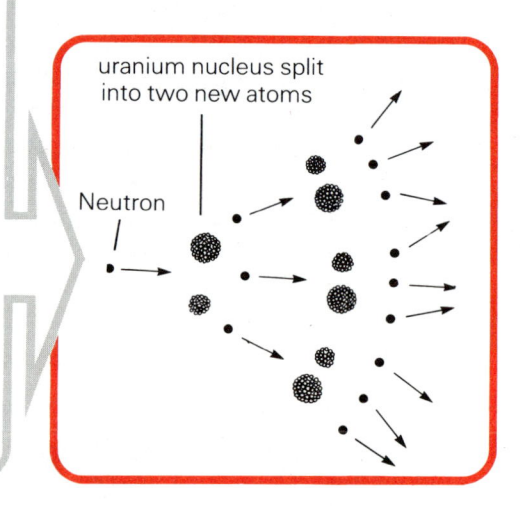

Controlled nuclear fission

Thankfully, the chain reaction in uranium-235 can be controlled to produce a steady supply of heat energy. This is done by absorbing roughly two out of the three neutrons produced by the fission. Only one neutron from a fission produces another fission and so on. This is a diagram of a nuclear reactor, where this process takes place.

Rods of uranium are held in a graphite block. Rods made of boron control the reaction by absorbing neutrons. The reactor is started by slowly taking the boron rods out. In an emergency, the boron rods can be dropped into the reactor to switch it off in less than a second!

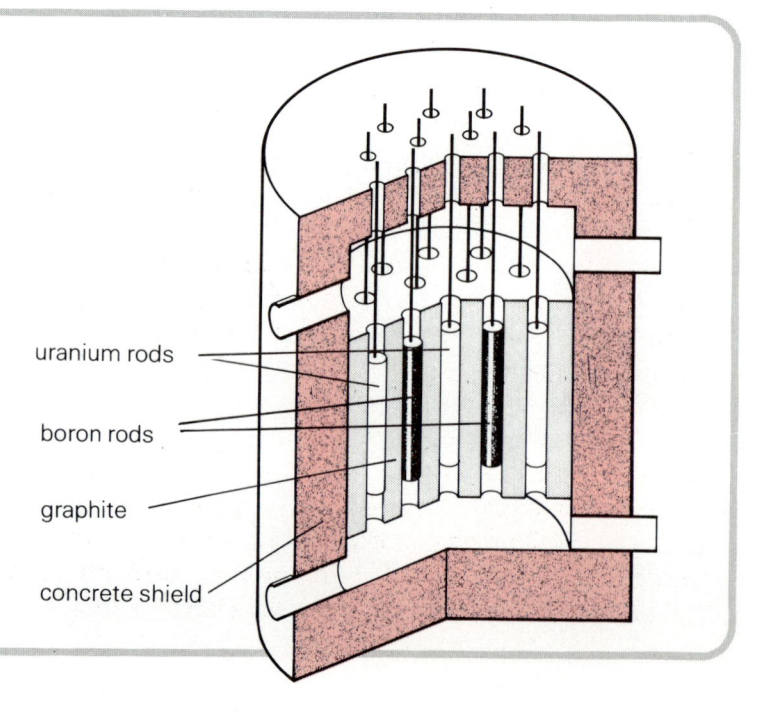

Nuclear power

The fission reactor gets very hot! The heat is used to make steam.

The steam turns a turbine which drives the electricity generator.

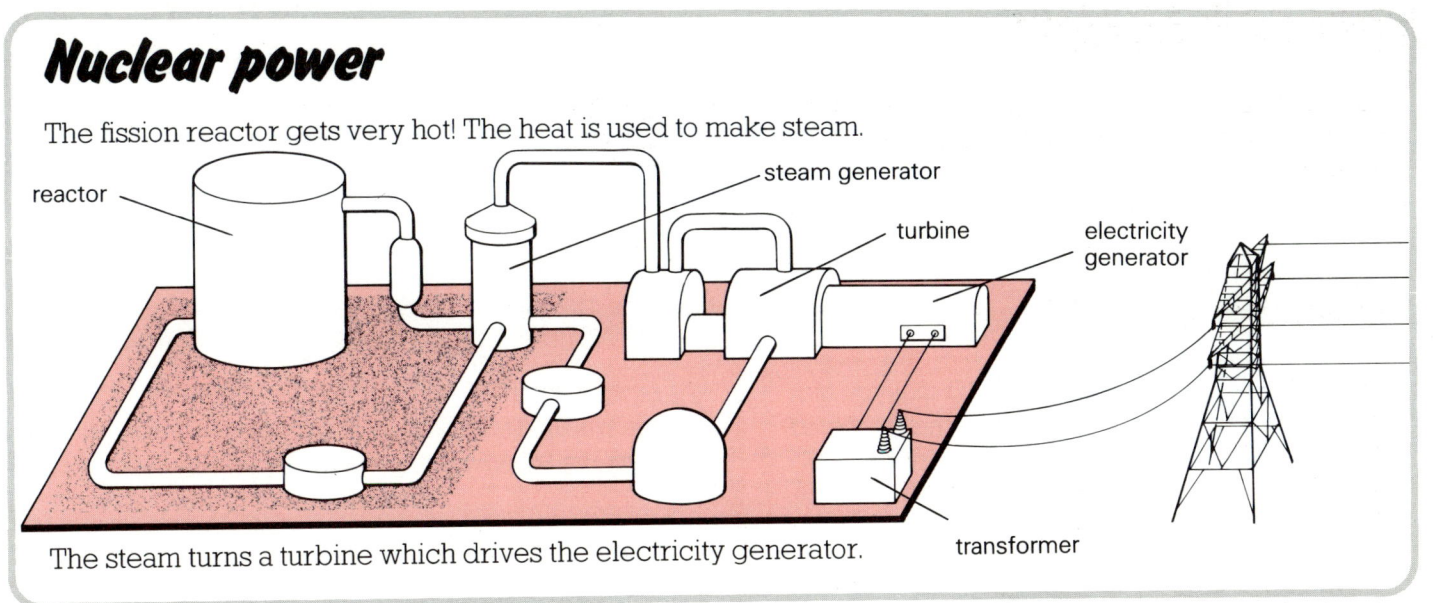

Nuclear fusion

There is another kind of nuclear reaction which produces energy. This is when two small nuclei join together to produce one new nucleus. The process is called **fusion**.

In nuclear fusion, the nuclei of two atoms of hydrogen join together. A helium nucleus is made. Most of the Sun's heat comes from the fusion of its hydrogen nuclei into helium.

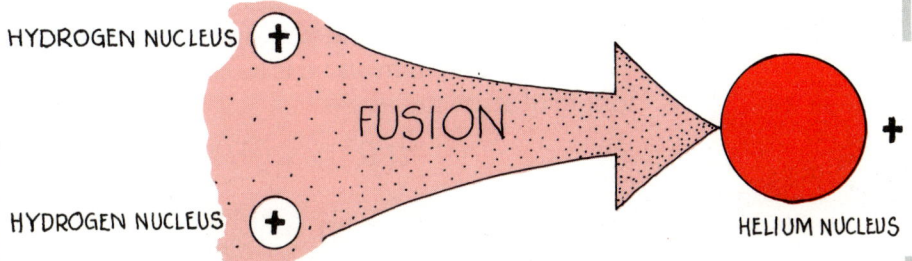

This sort of reaction gives much more energy than fission but it is very difficult to achieve. The fusing nuclei both have positive charges and repel each other. They can only be forced together at very high speeds. This means that they have to be heated to very high temperatures – somewhere in the region of 100 000 000°C.

Uncontrolled fusion has been achieved in hydrogen bombs. In these bombs a fission bomb is used to give the high temperatures needed to 'switch on' the fusion.

Controlled fusion is a most promising source of energy for the future. But it has not yet been achieved. The main difficulty is keeping the high temperature gas inside a container. The JET project at Culham in Oxfordshire is for research into the problems of fusion. Scientists working on the project there are hoping to trap the hot gas inside a magnetic 'bottle' for long enough for fusion to start.

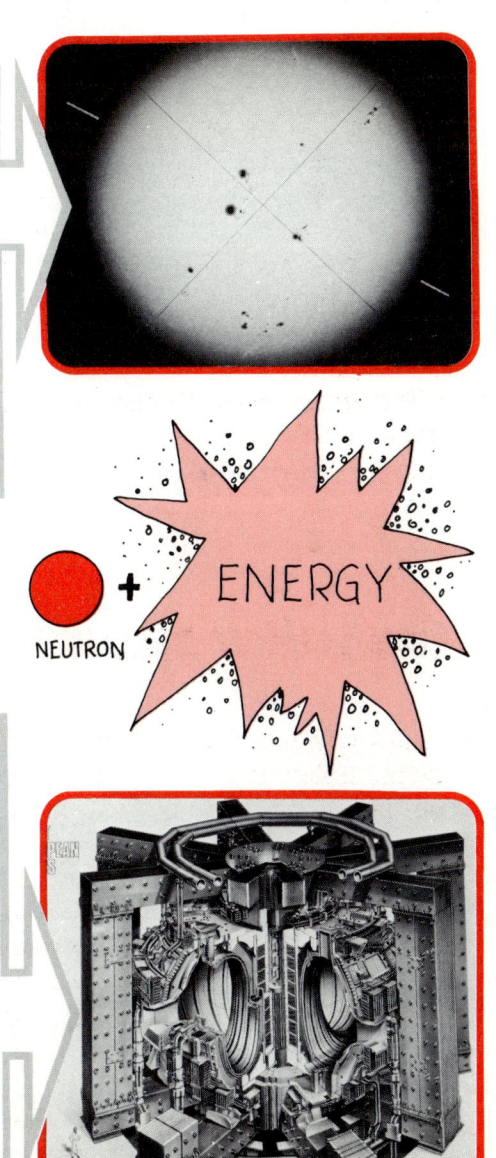

HOW BIG ARE MOLECULES?

Oil molecules are relatively large. This Activity shows you how to estimate their size. It works because oil spreads out on water into a layer only one molecule thick. The idea is to measure the volume of an oil drop and then let it spread out into a thin film on water. If you measure the area of the film, you can work out its depth using the volume of the original drop. The depth measurement is the length of one molecule.

You will need

- tray filled with water
- talcum powder
- olive oil
- small length of wire
- ruler • hand lens

1 Make a kink in the piece of wire so that you can collect a drop of oil on it. Measure the diameter of the drop with a hand lens and the ruler.

2 *Lightly* sprinkle talcum powder on the surface of the water. Touch the oil drop onto the surface near the middle of the tray.

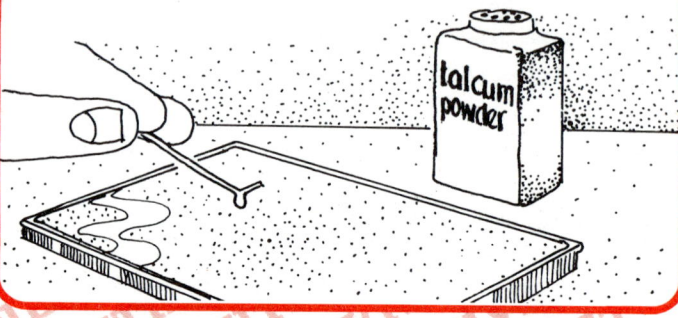

3 Measure the diameter of the oil film with the ruler.

4 The thickness of the film, is the length of the oil molecule. It can be calculated from the measurements.

Firstly, the volume of the oil drop is $\frac{4}{3}\pi r^3$ where r is its radius (½ of the diameter).

The volumes of the drop and the film are the same. The volume of the film is $\pi R^2 h$, where R is the radius of the film and h is its depth. h is the length of an oil molecule.

So: $\pi R^2 h = (4/3)\pi r^3$

therefore: $h = \dfrac{4\pi r^3}{3\pi R^2}$

cancelling: $h = \dfrac{4r^3}{3R^2}$

Before you start working out h – *guess* how big it will be!

Repeat the Activity using fresh water. Compare this measurement with your first.

Answer these questions

1. What is your value for h?
2. Do your measurements show that this method is accurate?
3. How could the Activity be improved?
4. Try other oils. Are the results the same?

THE SMOKE CELL

Molecules in a gas move around at high speeds all the time. They move so quickly, and there are so many of them, that they can push smoke particles about, even though smoke particles are much bigger. The movement of these smoke particles is called **Brownian motion**. It is so small that you need a microscope to see it. Before you look, you need to trap some smoke in a small clear box called a **smoke cell**.

You will need
- smoke cell
- microscope
- cover slip
- paper and wax straw (*not* a plastic one)
- power supply
- matches

1 Place the smoke cell on the stage of the microscope. Connect the power supply to the lamp. Adjust the voltage. Switch on.

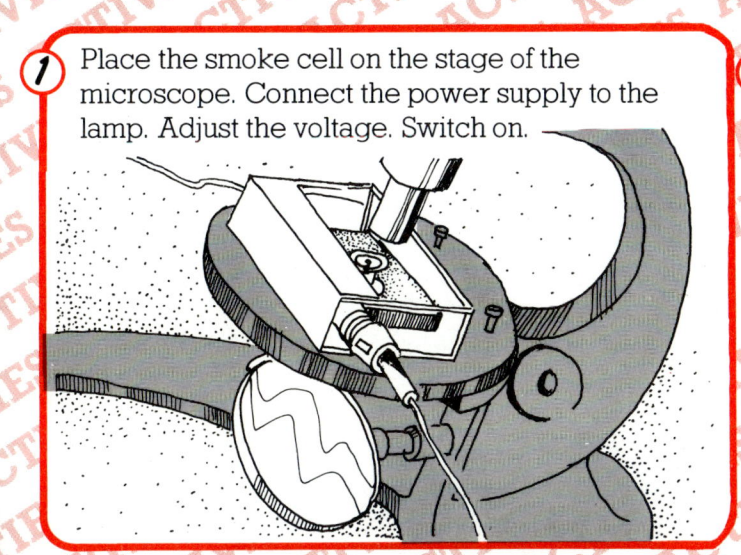

2 Light the straw at the end. Allow it to burn. Hold it as in the diagram, so that the smoke comes down the inside of the straw and goes into the smoke cell. Trap the smoke in the cell with the cover slip.

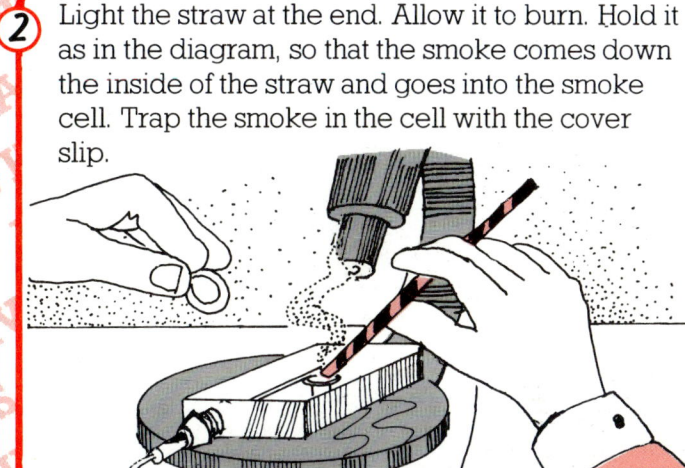

3 Use the lowest magnification. Lower the microscope until the objective lens nearly touches the cover slip. Look from the side while you do this. Don't lower the objective lens too far or you will damage the cover slip.

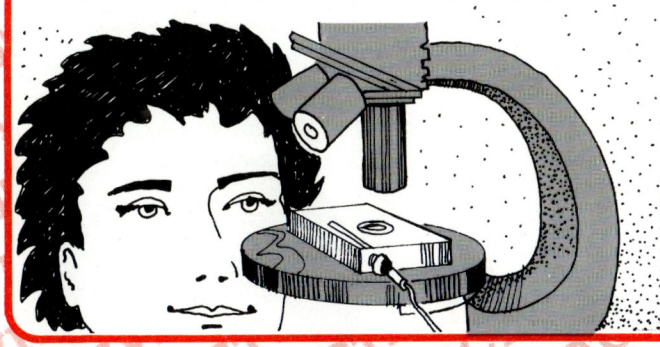

4 Look through the microscope and slowly raise the objective lens until you can see the tiny smoke particles. Describe their movement.

Answer these questions
1. Why do you need a lamp for the smoke cell? Try it without.
2. Explain what is happening in the smoke cell. What makes the smoke particles move about in the way they do?
3. What do you think would happen if this experiment was performed on a much hotter day?
4. What would you expect to see if you left the smoke cell exactly as it is for a week and then looked at it again. Try it!

105

INVESTIGATING SURFACE TENSION

In the first part of the Activity you are going to look at the **surface tension** in soap films. In the second part you are going to investigate **capillary action**.

You will need
- filter funnel
- candle
- thin wire
- thread
- soap solution
- two microscope slides
- a match
- a rubber band
- petri dish
- clean water

1 Blow a bubble on the filter funnel.

2 Take the funnel out of your mouth. Point the narrow end at the lighted candle. What happens to the candle flame? Can you explain why? What happens to the bubble? Why does it happen?

3 Make a small frame out of thin wire. Fasten a thread loosely across it. Dip the whole frame in soap solution. Break one side of the film by poking your finger through. What happens to the thread? Can you explain why?

4 Repeat step 2, but this time tie the thread so that there is a loop in it. What do you think will happen if you break the soap film inside the loop? Try it!

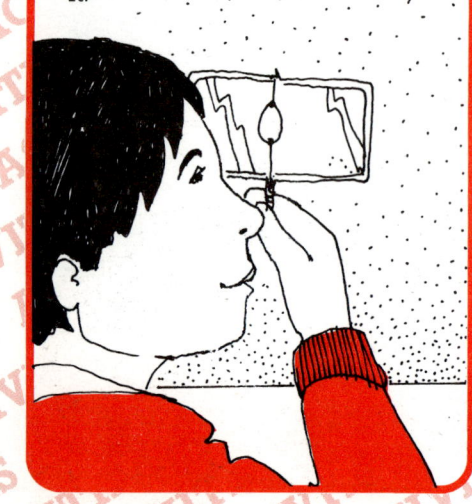

5 Join the microscope slides together with the elastic band, but keep them wedged apart at one side with part of the match. Lower the slides into some pure water in the petri dish.
   Can you see any capillary action? Draw what you see.

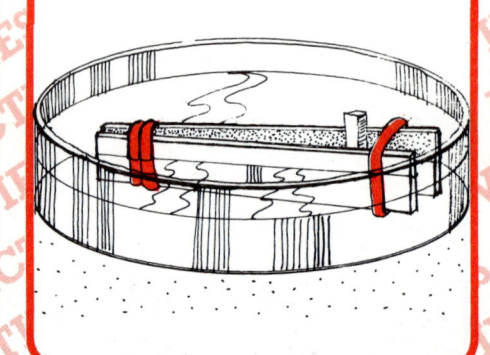

6 Add some detergent to the water in the petri dish. What happens? Can you explain it?

NUCLEAR POWER – FOR OR AGAINST?

This is a different kind of Activity!
Nuclear power presents many problems, but it also has many advantages. This Activity gives you a chance to think about the good and bad points of nuclear power. You'll then have a chance to present the case for or against.

Find out as much as you can about nuclear power. Add points of your own to these lists. Study the completed lists. Choose for or against, then either
(a) plan a video for TV to make all of your points as persuasively as possible. The programme should be 30 minutes long. You must say exactly how much time each section of the programme will take.
Or
(b) debate your side of the argument at a mock public inquiry into a proposal to build a nuclear power station near your town.

For

1. We cannot maintain our present standard of living, and improve on it, indefinitely. We will run out of fossil fuels (oil, gas, coal) sooner or later. If we wish to continue to enjoy the benefits of civilised living then we need to develop nuclear energy sources.
2. Nuclear power already supplies about 12% of our electricity. It is tried and tested and it has a tremendous safety record. It is much safer to produce nuclear power than to mine coal.
3. The United Kingdom leads the world in some aspects of nuclear power technology. The expertise of this technology can be sold to other nations. It is an important new industry which will create more jobs.
4. The disposal of the waste products is safe and does far less harm than leaks of oil from tankers.
5. Fossil fuels should be saved for the extraction of dyes, drugs, and other useful chemicals.

Against

1. We don't need to develop any new sources of power. By careful conservation of energy we can make do with the reserves of coal and oil that we have.
2. Apart from fossil fuels, there are other alternatives to nuclear power. What about tidal, wave, wind and solar power?
3. There is always the risk of a major disaster if the reactor got out of control. It seems unthinkable, but it might happen. What about precautions against a terrorist attack? As the number of power stations increases, can the security system cope?
4. Where does all of that radioactive waste go to? It is dangerous for a long time because of its long half-life. Aren't we storing up problems for future generations?
5. Nuclear power stations are very expensive. Surely alternative sources would be much more economical?

EXERCISES

1 Copy this out. Fill in the missing words.

The .... theory suggests that the .... in all things are constantly .... Those in a solid merely .... about a set position but those in a .... move rapidly to all parts of the container.

Choose from: moving, vibrate, gas, kinetic, molecules.

2 Explain how the walls of a house are built to prevent water getting into the house.

3 This little paper boat is on the surface of some clean water in a dish. Can you guess what will happen if one drop of washing-up liquid is placed on the notch? Try it and make a report on your findings!

4 What is surface tension? What does washing powder do to the surface tension of water? How does this help clean the washing? Explain how using shampoo makes it easy to wash your hair.

5 What is thermionic emission? Is it similar to evaporation?

6 List the three kinds of radioactivity. Which one is the most penetrating? Which one carries a positive charge? Which one is really a stream of fast moving electrons?

7 Explain what is meant by half-life. The activity of a piece of radioactive metal is 3200 counts per minute. What will it be in 10 days' time if the half-life is 2 days?

8 These readings show the radioactivity of some radon-220 gas.

Time/seconds	Counts/seconds
0	5003
20	3851
40	2982
80	1750
100	1343
130	900
200	356

(a) Plot out a graph of counts per second against time and draw in the curve.
(b) Work out the half-life of radon-220 from the graph.
(c) What was the count-rate at 60 s?
(d) Predict the count-rate at 400 s
(e) When was the count-rate exactly 1000 counts per second?
(f) Does each radioactivity count come only from the radon-220? How could each count rate be made more accurate?

9 Pair up these phrases. Copy out the complete sentences.

Capillary action is	the breaking up of large atomic nuclei by neutrons.
The meniscus is	the joining together of small atomic nuclei at high temperatures.
A cathode ray is	cannot get through paper.
An alpha particle	prevented in a house by a damp-proof course.
Nuclear fusion is	the name given to the 'skin' on water.
Nuclear fission is	a stream of electrons.

10 What is a chain reaction? Can you think of a way of explaining a chain reaction that would show the idea clearly to the rest of your class?

11 What is the source of the Sun's vast amount of energy? What has it got to do with what is going on at the JET project?

12 Draw a diagram of a nuclear reactor. Label the parts and say briefly how it works. How is the heat energy it makes taken away to be used?

7 ELECTROMAGNETIC WAVES

WHAT ARE THEY?

Light waves are a tiny part of a large range of waves known as the **electromagnetic spectrum**. The waves in this spectrum have different wavelengths and properties and are produced in different ways. They are all electromagnetic radiations, though. This chapter is all about these radiations.

LIGHT AND OTHER WAVES

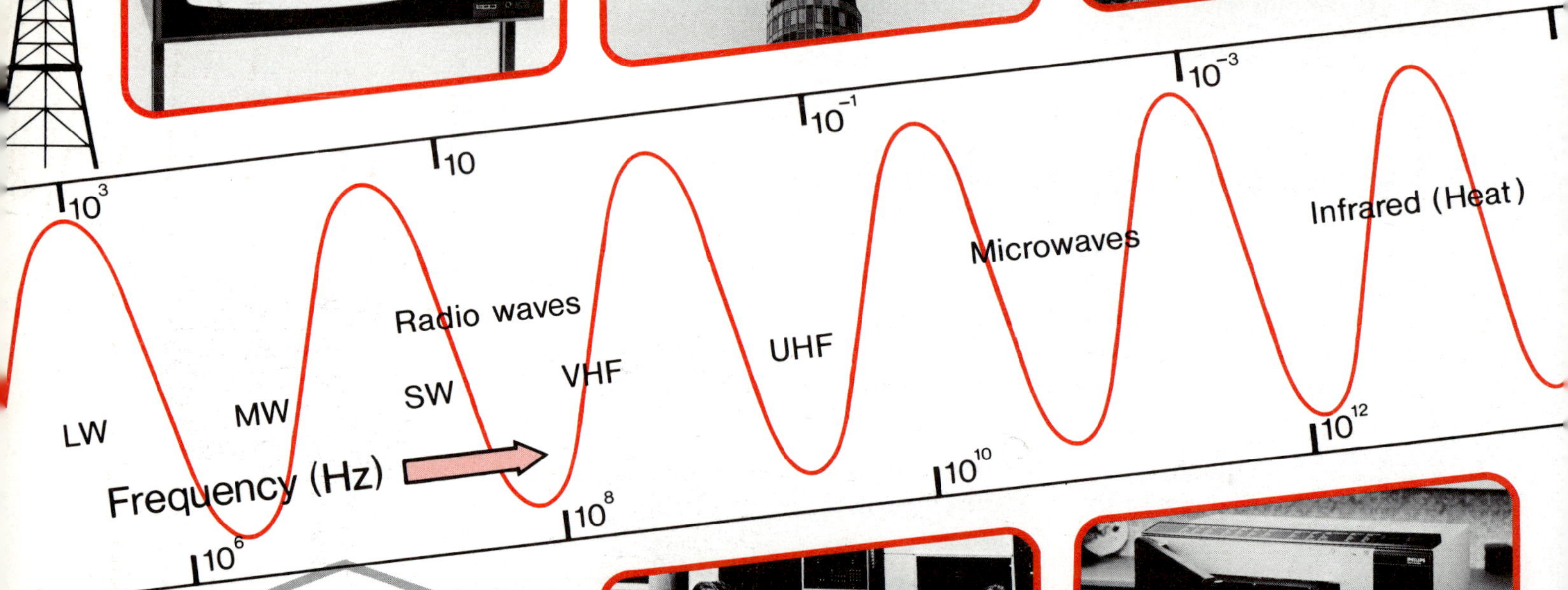

10^3 10 10^{-1} 10^{-3}

Infrared (Heat)

Microwaves

Radio waves

LW MW SW VHF UHF

Frequency (Hz)

10^6 10^8 10^{10} 10^{12}

At this end of the spectrum the radiations are very low in energy. They are not harmful to living things.

110

All the waves in the electromagnetic spectrum have the same velocity. They all travel at 300 000 000 m/s!

These waves obey the wave equation (Remember page 38). So, because the velocity is the same for all the waves, it follows that the shorter the wave length of the radiation, the higher the frequency.

RADIOACTIVE

At this end of the spectrum the radiations are very high in energy. They are very penetrating and harmful to living things.

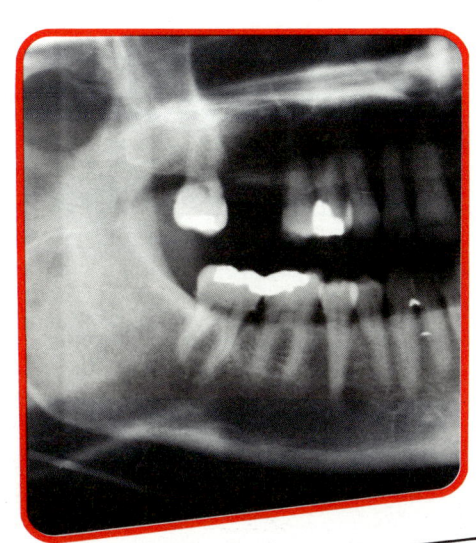

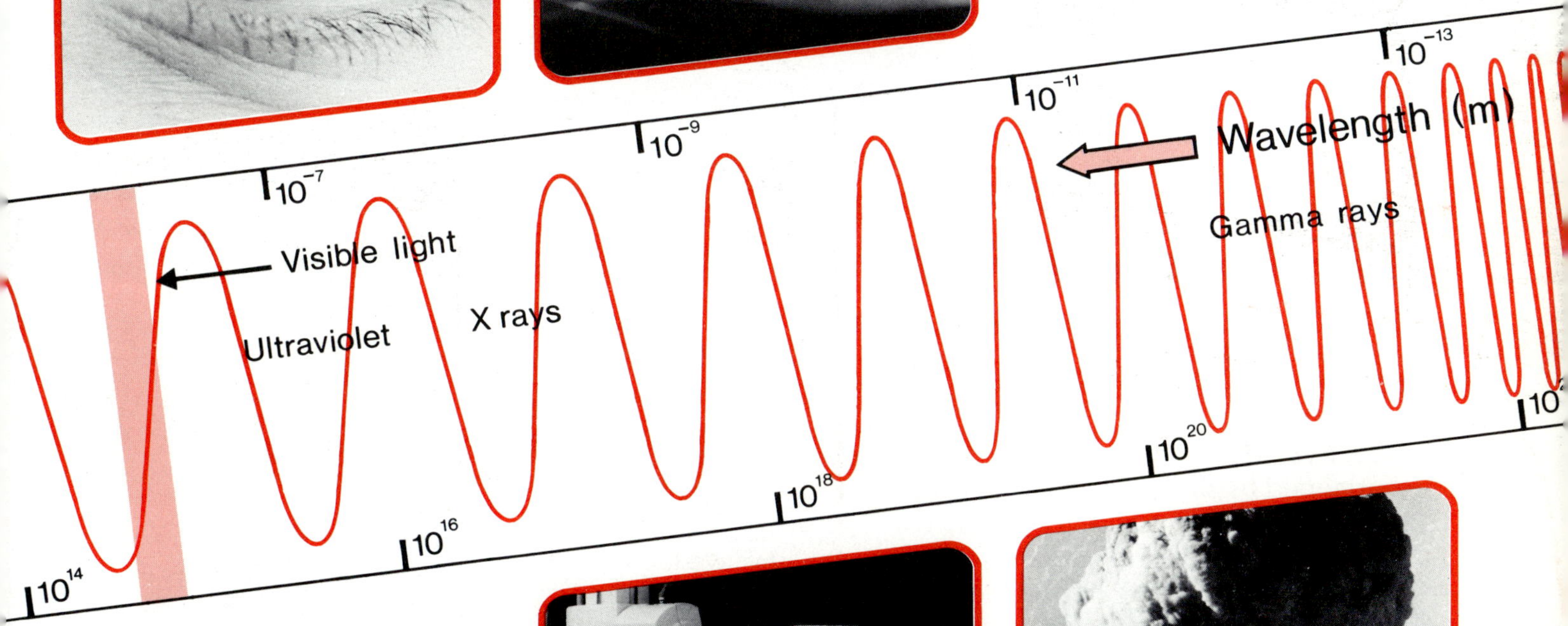

Wavelength (m)

Gamma rays

Visible light

Ultraviolet   X rays

10^{-7} 10^{-9} 10^{-11} 10^{-13}

10^{14} 10^{16} 10^{18} 10^{20}

Visible light is only a small section of the electromagnetic spectrum. This light, together with infrared radiation, is the only part of the spectrum your body can sense. What would it be like if you could sense the rest as well?

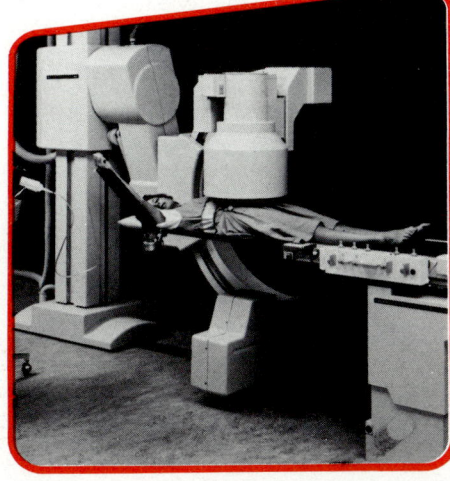

111

COLOUR

The spectrum

We call ordinary sunlight and the light from light bulbs white light. Although it appears white it's a mixture of different colours. Each colour has its own wavelength.

If you pass any white light through a prism you will see the colours. The prism refracts the different colours by slightly different amounts. So, they spread out. This is called **dispersion**. The diagram shows you what happens.

These colours are known as the spectrum

You can try this for yourself as part of the Activity on page 120. Red light is refracted least. It has the longest wavelength, about 7×10^{-7} m. Violet light is refracted most. It has the shortest wavelength, about 4×10^{-7} m. Seven colours have been named in the **spectrum**. You may find it hard to distinguish them. This is because wavelengths gradually change from one end of the spectrum to the other. There aren't any clear boundaries between colours.

Mixed coloured lights

Red, blue and green are known as the **primary** colours. These are coloured lights that cannot be made by mixing lights of other colours together. (This is not the same as mixing together paints. More about that later.)

Look at the back of this book. Red, blue and green lights can be overlapped by shining each one from its own projector onto a screen. This makes different colours.

The retina in your eye contains cells which are sensitive to different colours. It is no coincidence that there are three different kinds of cell. There are those that are particularly sensitive to red and those sensitive to blue. There is a third group sensitive to green. When all three kinds of cell are stimulated, you see white. Yellow stimulates the red and green together so that you see yellow and so on. Other mixed colours will stimulate the three sorts of cell by different amounts.

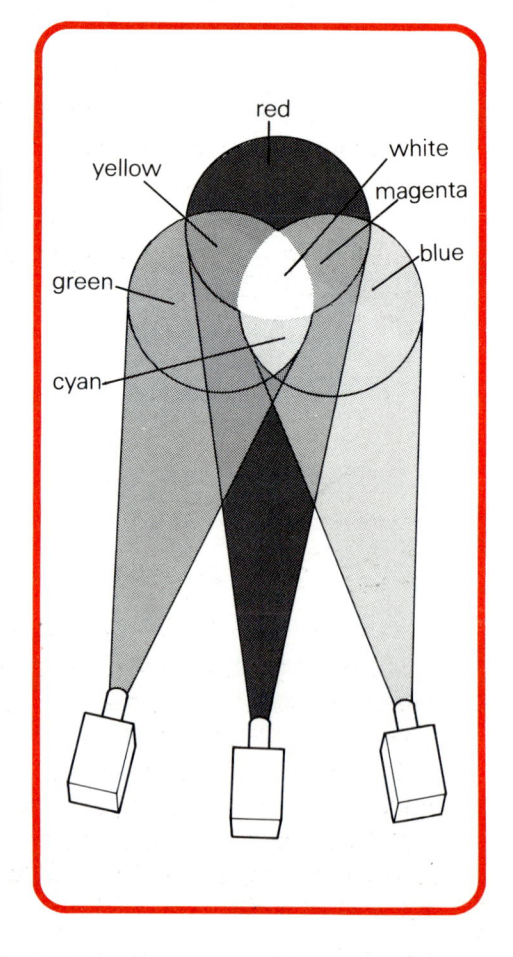

112

Colour television

In a colour TV tube there are 3 electron guns. They fire cathode rays at tiny dots on the screen. These dots show up blue, red or green when the rays hit them. Each gun fires at its own particular colour. The light from the dots mixes to make all the colours that you can see on the screen.

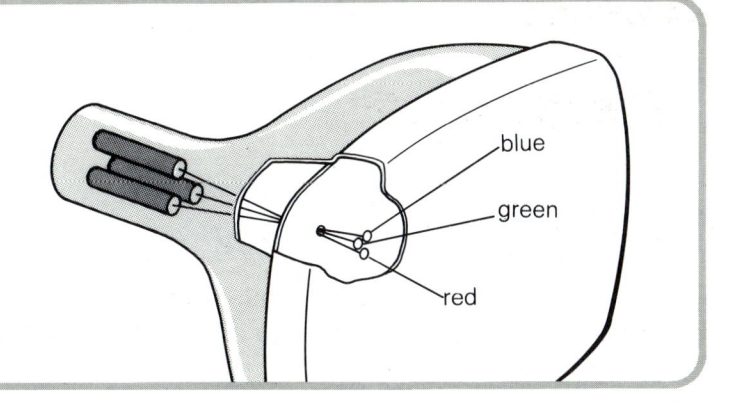

Pigments

Coloured objects contain coloured chemicals called **pigments**. All pigments reflect their own colour but absorb all others. So, a cricket ball looks red because its pigments absorb all of the colours of the spectrum except red. The red is reflected into your eye when you look at the ball.

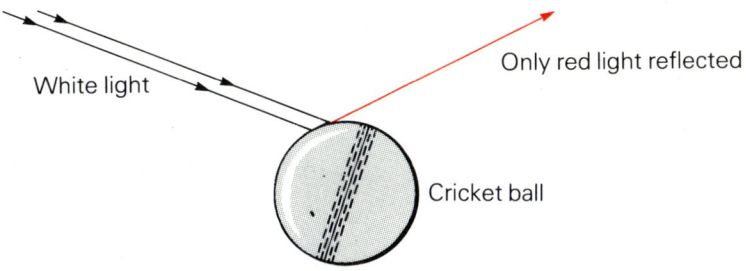

A white hockey ball reflects all of the colours of the spectrum. A black squash ball absorbs them all. It reflects hardly any light at all. A yellow T-shirt looks yellow because its pigments absorb blue light and reflect red and green.

Mixing paints

What happens when you mix yellow and blue paints? How do you end up with green? This is what happens. The pigments in yellow and blue paint together absorb all colours except green. This is the colour that is reflected. This is colour mixing by **subtraction**. The colours are *taken out* by the pigments.

Artificial light

If you look at something coloured under a street light you may not see its true colour. For example, a blue coat under the light from a yellow street-light looks black.

It works like this. The red and green coloured lights (in the 'yellow' light) are absorbed by the coat. No light is reflected. The coat can only reflect blue. But there is no blue to reflect! So, the coat looks black.

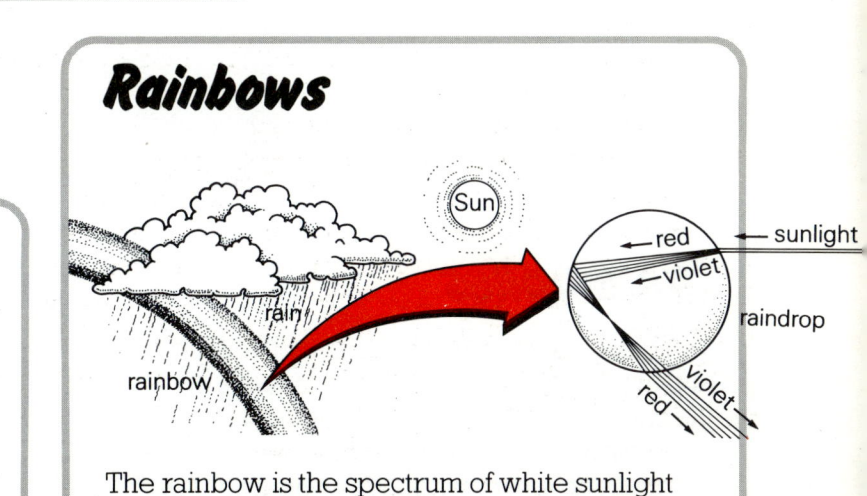

Rainbows

The rainbow is the spectrum of white sunlight shown on a grand scale. The prism in this case is millions of raindrops. Each raindrop disperses the white sunlight into its separate colours.

113

INFRARED RADIATION

What is it?

When you warm your hands by a fire you are actually feeling the infrared radiation given off by the flames. All things give out infrared radiation. The warmer they are the more they give out. So you are giving infrared out – even this book is giving out infrared!

You have already learned about the movement of heat energy by conduction and convection. (Look at page 80 if you need a reminder.) Infrared radiation is the third way that heat energy can move from place to place. The radiation of heat energy is usually known simply as **radiation**. Some surfaces are better than others at giving out radiation. Can you guess which ones are good at giving out radiation and which ones are good at absorbing it? The Activity on page 122 will help you to decide.

This person is giving out more infra red than usual

Infra red photographs

Most people have had their photograph taken in ordinary visible light. But did you know that you can also take photographs with infrared? There are special films that are sensitive to infrared but ignore visible light. This can be very useful. It give a way of sorting out hot things from cold without taking their temperatures. Two identical objects, one hot and one cold, may *appear* the same to look at. But infrared pictures of them will be very different.
Doctors sometimes use infrared photos of their patients to diagnose blood circulation problems or cancers.

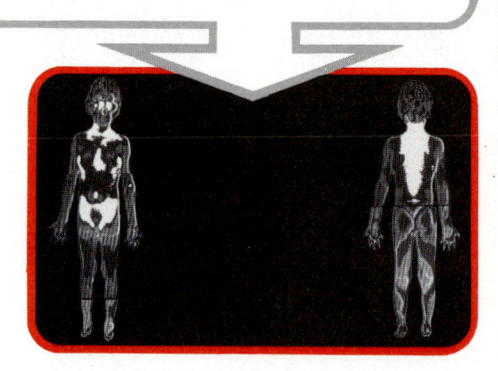

This is an infrared photograph of the UK.

Thermopiles

You can use a **thermopile** to detect very small amounts of infrared. A thermopile is lots of thermocouples (see page 79) joined together. A large current is made when all these thermocouples are warmed by infrared. This makes it more sensitive to tiny amounts of infrared than one thermocouple would be. The cone directs the radiation onto the thermopile.

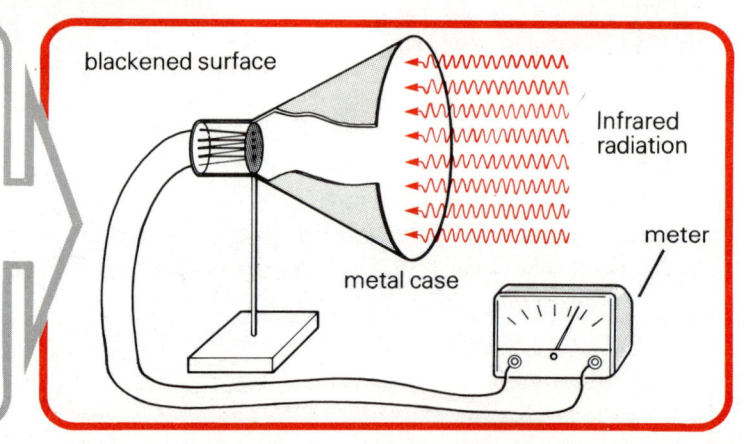

Why is it cold on clear nights?

Astronomers have a problem! Star gazing is only possible on a clear night. But clear nights are always cool. It is warmer on cloudy nights, but then you can't see the stars!

The cause of this problem is radiation.

On a clear night heat is radiated into space. The ground gets much colder.

On a cloudy night the heat is reflected back by the clouds.

Vacuum flasks

Vacuum flasks keep warm drinks warm or cold drinks cold. The vacuum in the walls stops heat energy moving in or out by conduction or convection. The glass walls of the vacuum flask are silvered like a mirror. This reduces heat loss even further. The shiny surface does not radiate heat very well. You can find out how well a vacuum flask works in the Activity on page 91

The silvery sides of a vacuum flask reduce heat loss

How do greenhouses work?

Have you ever noticed how hot it gets inside a greenhouse on a sunny day? Greenhouses trap the Sun's energy.

The Sun is very hot. This means that it gives out infrared radiation with a short wavelength. Short wavelength infrared can go straight through the glass into the greenhouse. This warms up the things inside. They emit their own infrared. They are a lot cooler than the Sun, so their infrared has a long wavelength. This can't get through the glass. It is reflected back into the greenhouse. This is known as the **greenhouse effect**.

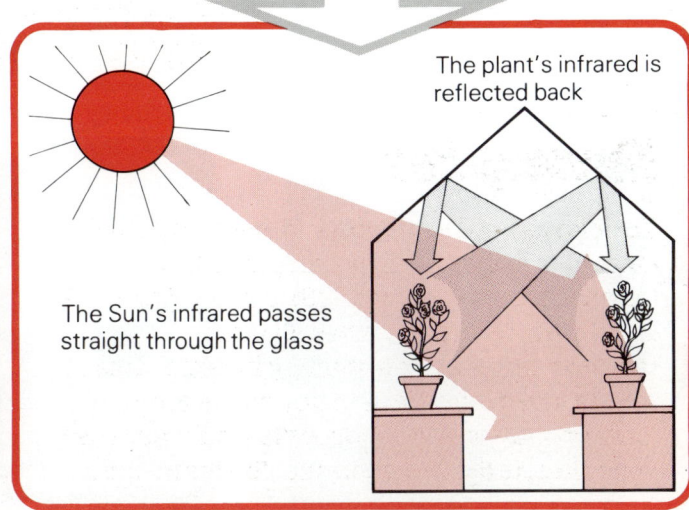

The plant's infrared is reflected back

The Sun's infrared passes straight through the glass

ULTRAVIOLET

Ultraviolet waves are given out by atoms of very hot things. These waves have *shorter* wavelengths than the violet light you can see. That's why they're called *ultra*violet. They have more *energy* than violet light, though. You can't see ultraviolet. You can only see what it does.

Ultraviolet and you

Ultraviolet gives you a sun-tan. When your skin is exposed to ultraviolet from the Sun or a lamp it makes a brown pigment called **melanin**. Your skin can also use ultraviolet to make vitamin D. But too much ultraviolet is harmful. It can damage your eyes and your skin.

Fluorescence

Some substances absorb ultraviolet and then release the energy as a visible light glow. This is **fluorescence**. The inside of a fluorescent light tube is coated with a special powder. The powder absorbs ultraviolet and gives out white light. The ultraviolet is made by the electricity as it passes through the tube.

Fluorescence caused by ultraviolet has many other uses.

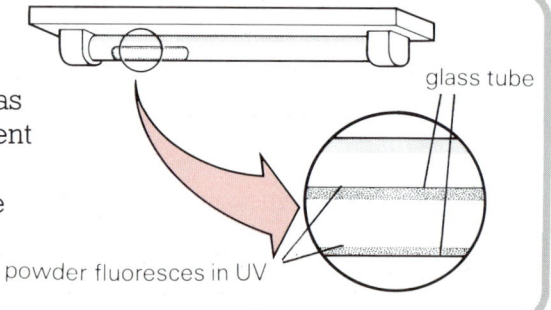

glass tube

powder fluoresces in UV

Some soap powders contain a chemical which fluoresces in ultraviolet. The chemical sticks to your shirt or blouse when it is washed. So, when you walk about in sunlight the fluorescence makes it look 'whiter than white'.

Have you ever seen the purple disco light which makes your clothing glow and your teeth shine? It is actually a very weak ultraviolet lamp. You can't see the ultraviolet radiation. You can only see the fluorescence.

Coloured clothing fades in sunlight. This is because the ultraviolet breaks up the molecules of the pigments. This is also why some wines and beers are sold in dark brown bottles.

X-RAYS

What are they?

X-rays are short-wavelength electromagnetic radiation. Their wavelengths are even shorter than ultraviolet. They have more energy than ultraviolet too.

Certain X-rays pass through flesh very easily but are stopped by bones. As X-rays affect photographic film, doctors can use them to find out about fractures and diseases.

Have you ever had an X-ray?

How are X-rays made?

X-rays are made when high energy electrons hit a metal target. A special X-ray tube is used.
Electrons from the filament are speeded up by a very high voltage. They hit the tungsten target. About 1% of their energy is converted into X-rays. The rest becomes heat. This has to be conducted away by the cooling liquid.

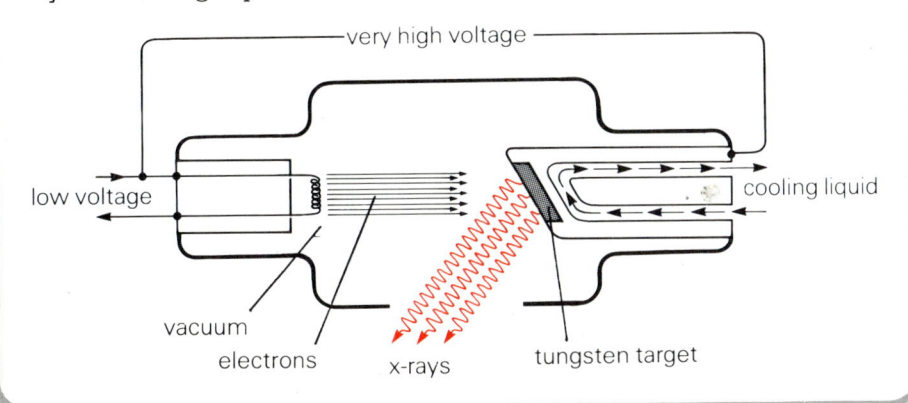

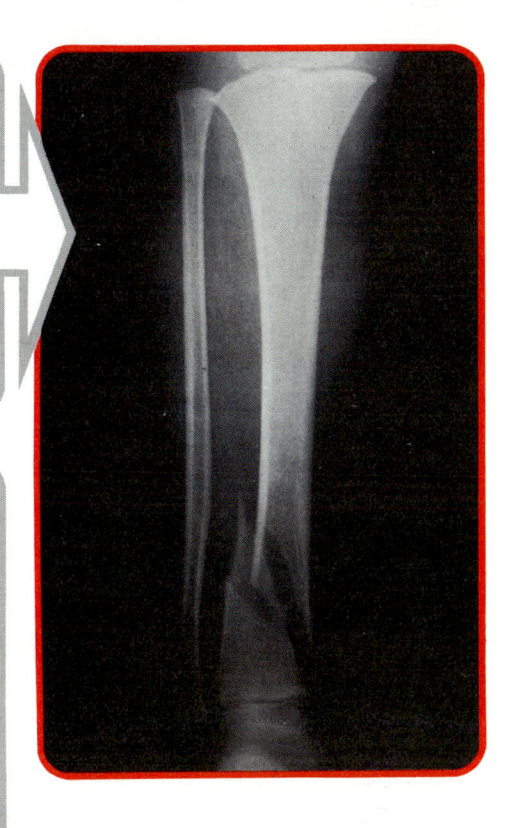

GAMMA RAYS

Gamma rays contain more energy than any of the other radiations from the electromagnetic spectrum. They have very small wavelengths. They are the most penetrating. They come from certain forms of radioactive decay. They may also come to the Earth from the Sun or Space. All of the particles and waves that reach us from stars are called **cosmic rays**. Some cosmic rays are gamma rays. Gamma rays pass through human bodies very easily. They can do great damage to the cells. It is very difficult to get protection from gamma rays. In fact, a suit made with lead several centimetres thick won't stop them completely. The best protection is to keep away from radioactive sources!

RADIO

Radio waves

Radio waves have the longest wavelengths in the electromagnetic spectrum. There are many kinds of radio waves. They are grouped together in **frequency bands**. These bands are useful in different ways for different things.

All radio waves come from electrons which are made to **oscillate** (that means move backwards and forwards) quickly. The waves may be sent out and received by aerials. Tiny currents called **electronic signals** can be added to the radio waves. In this way information such as TV programmes or telephone messages can be sent great distances. The extra electronic signals come from TV cameras, microphones and so on.

Long, medium and short waves

A lot of radio programmes are transmitted by long waves or medium waves. These waves are particularly useful because they will reflect from layers of the Earth's atmosphere. So, even though the Earth curves, signals can be sent over great distances.

Short wave radio signals are also reflected by layers of the Earth's atmosphere. But they are *also* reflected by the Earth's surface. Because of this, it's possible to send messages right around the Earth.

VHF and UHF

Very High Frequency (VHF) and Ultra High Frequency (UHF), have very short wavelengths (remember page 41). They are used for stereo radio and TV broadcasts. They give very good quality transmissions but they are not reflected by the atmosphere. There has to be a straight line between the transmitter and your aerial.

118

MICROWAVES

Microwaves have wavelengths shorter than 70 cm. They have many uses – mainly in communications. The British Telecom Tower beams out microwaves. They are passed on by relay towers all over the country.

Satellites

Microwaves are used in satellite communication. Signals can be sent to satellites, given a boost, and then returned to another part of the Earth.

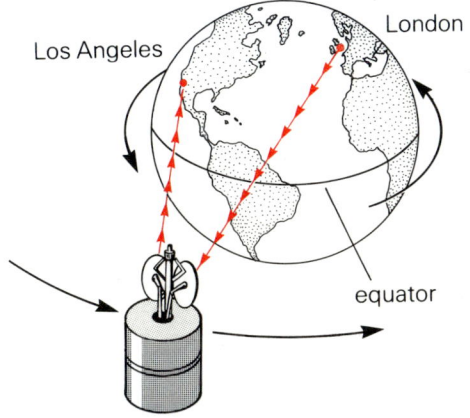

A **geostationary satellite** stays in the same place above the Earth's surface. It has to be directly over the equator and orbits once every 24 hours to do this.

Communications satellite

Microwave ovens

Microwaves are used in cooking. They are easily absorbed by food, especially the water molecules in the food. The food gets hot and cooks very quickly.

Radiotelescopes

Many radio waves come from outer space. They can tell us lots of things about the Universe. The study of these radio waves is called **radio astronomy**.
Jodrell Bank is Britain's most famous radio telescope.

Radar

Microwaves are used in **radar**. The waves are sent out and the reflections from solid objects are picked up when they come back. The distance to the object reflecting the waves, in this case the rocky shore, can be worked out and evasive action taken!

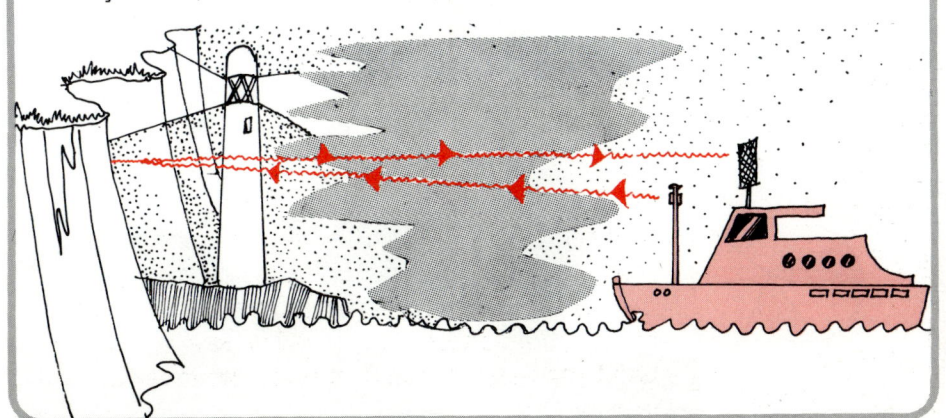

119

REVERSING THE RAINBOW

If you pass white light through a prism or a raindrop you can make a spectrum of colours. But you can also do the reverse – make white light from a rainbow.

You will need:
- two prisms
- ray box
- slit
- felt tip pens
- thin string
- scissors
- pair of compasses
- cardboard
- screen

1. Set out the ray box, screen and one prism to give the best possible spectrum on the screen.

2. Put the second prism in between the first one and the screen. Point it in the opposite direction to the first one. What do you see on the screen?

3. Alter the position of the second prism and see what effect it has. Does the angle of the ray from the ray box make any difference?

4. This is how to make a Newton's disc. Cut out a circle, about 10 cm in diameter from the stiff cardboard.

5. Divide one side of it into seven segments. Colour them with the colours from the spectrum.

6. Put two holes in the disc. Thread a loop of string through. Now make it spin! What can you see on the coloured side? Can you explain it?

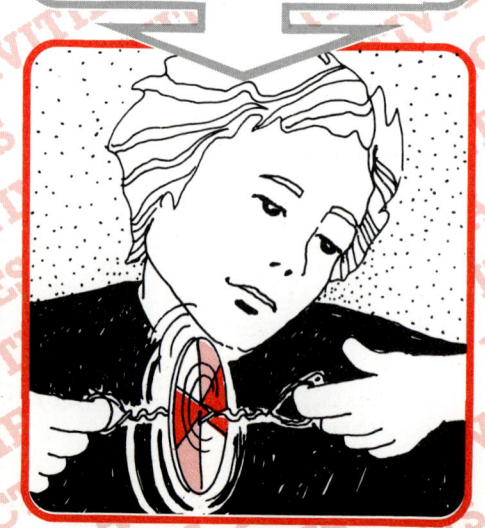

WHAT DO FILTERS DO?

Filters are thin transparent pieces of coloured plastic. They allow only certain wavelengths from the spectrum of white light to pass through. In the first part of this Activity you can find out more about the dispersion of white light. You can then use this dispersed light to find out about filters.

You will need
- paper
- prism
- ray box and slit
- power pack
- white paper
- different filters
- different felt tip pens

Copy this table

| Filter colour | Predicted colour on the screen | Actual colour on the screen |

1 Arrange the ray box and prisms so that you get a good spectrum on the screen.

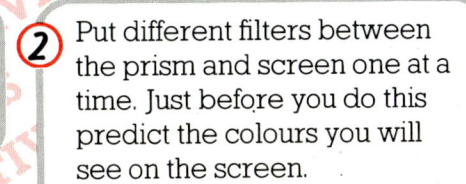

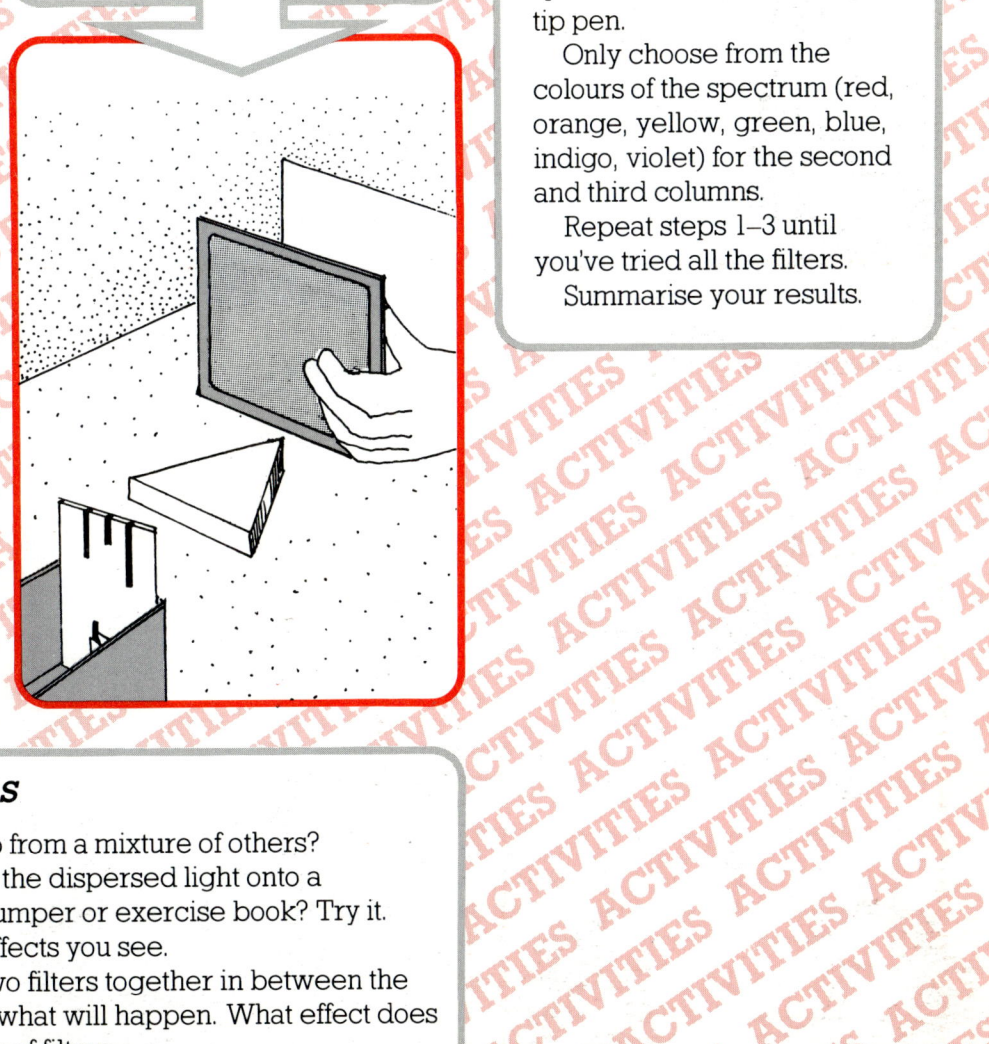

2 Put different filters between the prism and screen one at a time. Just before you do this predict the colours you will see on the screen.

3 Write your results in the table.
You may have trouble describing the colour of a filter. If you do, put a patch of the colour *most like it* in the space in the table. Use a felt tip pen.
Only choose from the colours of the spectrum (red, orange, yellow, green, blue, indigo, violet) for the second and third columns.
Repeat steps 1–3 until you've tried all the filters.
Summarise your results.

Answer these questions

1. Which colours are not made up from a mixture of others?
2. What happens when you shine the dispersed light onto a coloured screen such as your jumper or exercise book? Try it. Make a list of any interesting effects you see.
3. What happens when you put two filters together in between the prism and the screen? Predict what will happen. What effect does it have? Try other combinations of filters.

WHICH SURFACES RADIATE INFRA RED BEST?

In this Activity you will be able to find out which surfaces are good absorbers of infrared radiation and which surfaces are poor absorbers.

You will need
- two metal plates, one matt black, the other shiny silver
- wax
- two identical coins
- clamps and stands
- a radiant heater
- matt-black tin can
- shiny tin can
- lid for the tin cans
- two thermometers

Copy this table

Time/minutes	Temperature of matt black can /°C	Temperature of shiny can /°C

1. Stick one coin on the back of the matt-black plate with melted wax from the candle. Stick the other coin on the back of the shiny plate in the same way. Wait until the wax is solid.

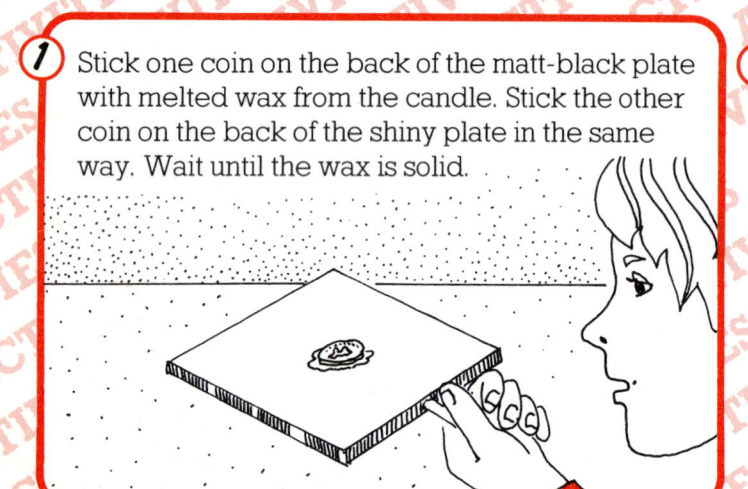

2. Clamp each plate to a stand. Make sure the shiny and matt surfaces face each other across a 30 cm gap. Put the radiant heater exactly in the middle, between the plates. Switch on.
   Predict which coin will fall off first.

3. Put the same amount of water in the two metal cans. Place a thermometer in each one. Put the lids on. Place the can on either side of the radiant heater. Make sure each can is the same distance from the heater.

4. Record the temperatures every five minutes in the table. Draw two line graphs, one for each beaker, on the same axes.

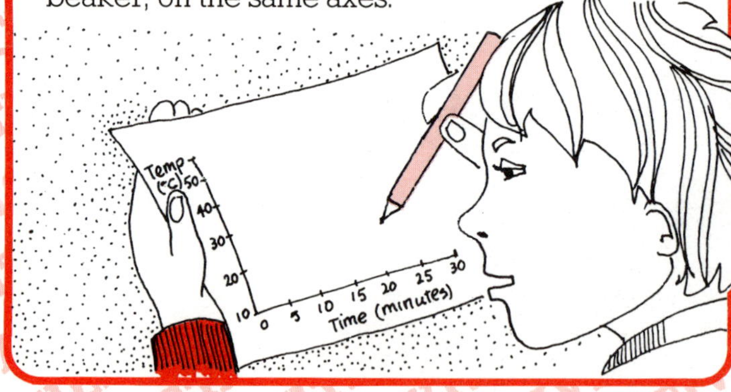

Answer these questions
1. Which can absorbed more heat energy?
2. Do the results of the two parts of the Activity agree?
3. How does the energy get from the heater to the cans? How does it get from the side of the can to the water?
4. Why is it important to make sure that both plates and both cans are the same distance from the heater?

WHICH SURFACES ABSORB INFRA RED BEST?

Some surfaces radiate infrared waves better than others. In this Activity you are going to find out which are good radiators and which are poor radiators.

You will need
- matt-black tin can
- shiny tin can
- 2 thermometers
- boiling water
- lids for the tin cans
- large can with one side painted matt black, the other side shiny
- thermopile
- galvanometer

Copy this table

Time/minutes	Temperature of matt black can /°C	Temperature of shiny can /°C

1. Pour the same amount of boiling water into the tin cans. Place the thermometers in. Put the lids on. Predict which one will be warmer after 20 minutes. Write the temperatures of the cans in your table. Start by taking readings every minute. After about 5 readings take the temperature every 5 minutes.

2. Draw two line graphs of your results on the same axes.

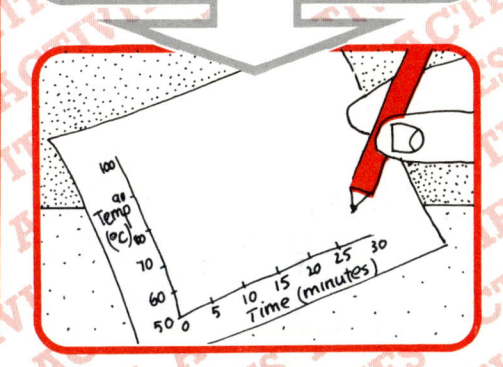

3. Fill the larger black and shiny can with boiling water.

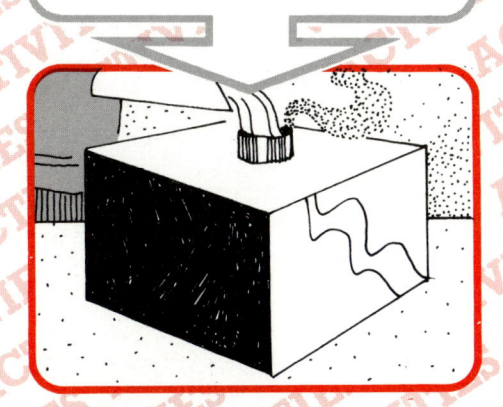

4. Connect the thermopile to the galvanometer. Point the thermopile towards each surface in turn. Before you take the readings guess which surface will give the higher reading. Which surface did?

Answer these questions

1. What does the thermopile do? Look at page 114.
2. How do these results compare with the results of the Activity on page 122? Summarise the results of these two Activities.
3. This marathon runner is wrapping himself in shiny foil blankets. Why? Why does the blanket work so well?

123

EXERCISES

1. Put these radiations in order of increasing wavelength: X-rays, infrared, radio waves, gamma waves, visible light.
   Which radiation from the electromagnetic spectrum is missing from this list?

2. Which radiations from the electromagnetic spectrum have these frequencies: 10^{13} Hz; 10^{16} Hz; 10^{20} Hz?
   How are these frequencies related to the wavelengths of the radiation?

4. The wavelength of violet light is 4×10^{-7} m and its velocity is 3×10^8 m/s. What is its frequency?

5. What is infrared radiation usually known as? How can infrared be detected? Why is it cold on a clear night, but warm on a cloudy night?

6. Some people claim that burning coal and oil at our present rate, will badly pollute the atmosphere with carbon dioxide. Carbon dioxide will trap radiation and we will suffer from the 'greenhouse effect'? What is the 'greenhouse effect'?
   What do you think will happen if the predictions are correct? Is it likely?

7. People sometimes remember the first letters of the colours in the rainbow with this trick
   Richard **O**f **Y**ork **G**ave **B**attle **I**n **V**ain
   Can you think up your own sentence using those first letters?

8. Why does a blue dress look blue? What colour would it look in red light? What about yellow light? How would it look in magenta light?

9. Copy this out. Fill in the gaps

   When you shine white light through a ____ it ____ the different colours by different amounts. This is called ____ The colours you can see are called the ____ Three of these colours ____ and ____ are called the ____ colours. You can mix these to make other colours. Red and ____ makes magenta, ____ and green makes yellow, green and blue make ____

   Choose from spectrum, red, green, prism, refracts, dispersion, blue, cyan, primary.

10. Make a list of the uses of radio waves.
    Which part of the electromagnetic spectrum has not yet been put to use?

11 Are these statements true or false? If they are false, explain why.
  (a) Suntans are caused by the ultraviolet in sunlight.
  (b) X-rays have no effect on the film in an ordinary camera.
  (c) Gamma rays have longer wavelengths than infrared.
  (d) Ultraviolet radiation contains more energy than infrared.
  (e) VHF broadcasts can be heard easily because the waves reflect from clouds in the sky.

12 A girl set up an experiment in the laboratory. She produced a spectrum of light from the Sun's light coming in through a hole in the blinds. It shone on a white screen. She painted the bulb of a thermometer matt black and held it in the dark just to the left of the red light in the spectrum. The temperature recorded by the thermometer went up steadily even though it was in the dark. When she held the thermometer to the right of the spectrum the temperature did not change. Explain her results. (You could try this experiment yourself!)

INDEX

absolute zero 87
acceleration 9, 15, 16
action and reaction 6
alpha particles 100
ammeter 63, 74, 75
   milliameter 72, 73
amplitude 41, 45
atoms 78, 94, 98–103, 116

battery 62, 63, 72
beta particles 100
bimetal strip 86, 89
binoculars 27
boiling
   point 85
   water 78, 79
Brownian motion 105

camera 31
   pinhole 32
capacitors 66,
capillary action 97, 106
cathode rays 99
Celsius scale 79
centre of gravity 8, 17
colour
   primary 112
   television 113
comets 12
compass 57
conduction 80, 91
constellations 13, 18
convection 81, 83, 90
cosmic rays 117
critical angle 35
current 62, 63, 65, 66, 72–75
   a.c./d.c. 62, 63, 71, 75

depolariser 72
diffraction 50
diffusion 95
diodes 66, 67
Doppler effect 47
double glazing 82
dynamo 62, 73

ear 46
Earth 10, 23, 57, 119
earthed 60, 61, 64, 65
echoes 43
eclipses 23
electric bell 58

electrolyte 72
electrostatic 60–61
electromagnets 58, 59
electrons 60, 61, 66, 98–99, 101
energy 40, 78, 79, 82, 102, 103
   kinetic 78, 94
evaporation 85, 95
expansion 86, 87
eye 30, 31, 111

fibre optics 27
filters 121
fluorescence 116
focal length 29, 30, 31
forces 4, 5, 6, 7, 96
freezer 85
frequency 40, 41, 43, 44, 45, 47, 118
friction 5, 6, 60
fridge 85
fuses 64, 75

galaxy 12, 13
gamma rays 100, 101, 117
gravity 4, 8, 9, 16, 17
greenhouse 115

half-life 100
hot air balloons 83

image 25, 28, 29, 30, 31, 36, 37
indicator lights 89
induction, magnetic 56, 57
   electrostatic 69
   electromagnetic 73
infrared 111, 114, 115, 122, 123
insulator 60, 80, 82, 91
integrated circuit 67
interference 50
ionisation 100

JET project 103
joulemeter 84

kaleidoscope 34
Kelvin scale 87
kinetic theory 94

latent heat 84, 85

lenses 28–31, 36, 37
levers 7
lightning 60, 61
loudspeaker 59

magnets 56, 57, 58, 63, 68, 73
magnifying glass 28, 36
meniscus 97
melting 78, 84, 88
meteors, meteorites 12, 19
microscope 29, 105
microwaves 110, 111, 119
Milky Way 12
mirages 26
mirrors 22, 25, 27, 33, 34
moments 7
motor 70
music 44, 45, 53
Moon 10, 11, 18, 23, 24

National Grid 64
neutrons 98, 102
newtons 5
nuclear energy 102, 103, 107
nucleus 60, 101, 103

octave 44
ohms 66
oscilloscope 62

periscope 25, 27, 34
pigments 113
planet 10, 11, 19
plug (13 A) 65, 75
pole, north or south 10, 56, 57, 68, 70, 71
potential difference 63, 74
potentiometer 66
pressure, air 85
prisms 27
protons 98, 101
projector 28, 29

radar 110, 119
radiation 91, 101, 110, 111, 114, 115, 117
radio 110, 118
radioactivity 100, 101, 102, 117
rainbow 113, 120
rectifier 66

reflection 11, 24, 25, 33, 49, 119
refraction 26, 35, 50
resistors 66, 74, 75
resonance 43

satellites 119
semiconductors 67
shadows 22
silicon chip 67
solar system 10
sound 42, 44, 45, 46, 59
speed 9, 14, 15, 41, 51
stars 12, 13, 18, 19
stability 8
stroboscope 48–50
Sun 10, 23, 24, 103, 115
surface tension 96, 106
synthesizers 45

telephones 59
telescopes 29, 37
temperature 78, 79, 83, 84, 86, 88, 91
thermionic emission 98, 99
thermocouple 79, 114
thermometers 79
thermopile 114
thermostat 83
transformer 64, 66
transistors 66, 67
tuning fork 44, 53

ultraviolet 111, 116

vacuum flasks 91, 115
velocity 9, 111
vocal cords 47
volts 63, 64, 74
voltmeter 63, 74

watts 75
wavelength 41, 48, 110–112, 115, 118

X-rays 117